REASON IN REVOLT

REASON IN REVOLT

DIALECTICAL PHILOSOPHY

AND

MODERN SCIENCE

Volume I

Ted Grant

&

Alan Woods

Algora Publishing

New York

ISBN: 0-87586-156-3 (softcover)
ISBN: 0-87586-157-1 (hardcover)

Library of Congress Cataloging-in-Publication Data: 2002014693

Grant, Ted.
  Reason in revolt : dialectical philosophy and modern science / Ted Grant, Alan Woods.
     p. cm.
Includes bibliographical references.
  ISBN 0-87586-157-1 (alk. paper) — ISBN 0-87586-156-3 (pbk. : alk. paper)
  1. Communism and science. 2. Dialectical materialism. I. Woods, Alan. II. Title.

HX541 .G736 2002
335.4'112—dc21
                        2002014693

Printed in the United States

TABLE of CONTENTS

Author's Preface to the North American Edition

Marxism and the United States

By Alan Woods

The publication of *Reason in Revolt* in the USA is a real red-letter day for the authors. The appearance of the book seven years ago was greeted with enthusiasm by many people, not only on the left, but by scientists and other people interested in philosophy and the latest scientific theories, such as chaos and complexity, which in many respects reflect a dialectical approach to nature. It is therefore highly appropriate that the American public should be acquainted with that extraordinary body of ideas called Marxism.

We are living in a period of profound transformations on a global scale. The exciting advances of the last decades in genetics and physics, the development of computer science, the Internet and other fields of science and technology are opening up new horizons for human development. As the boundaries of human knowledge are pushed forward, so the potential for human development reaches new and unsuspected levels. In this way, the foundations are being laid for a revolution that will put all past transformations in the shade."

The advances of science and technique can provide the basis for the eradication of the evils that plague humankind at the beginning of the 21st century. In the Bible it is written "for the poor have ye always with you." Yet now, for the first time in human history it is possible to abolish poverty. There is no need for any human being to be without a house, a job, a decent education. There is no need for seven million children to die each year of elementary diseases connected with the lack of clean water.

We have within our grasp the potential to give every man, woman and child on the planet a genuinely human condition of existence. But this colossal potential is far from being achieved. Alongside the most staggering advances of science and culture, we see the most appalling misery, ignorance and deprivation on a world scale. The phenomenon of globalization (which was predicted by Marx and Engels over 150 years ago), for two-thirds of humanity, means only the generalization of this misery, suffering and despair.

The present epoch is therefore confronted by a fundamental contradiction, the resolution of which will decide the fate of humankind. On the one hand, all the material means exist for creating a paradise in this world. On the other hand, our planet is being turned into a living hell for countless millions. Everywhere we look there is turmoil, conflict, wars, instability. The "invisible hand of the market" that was supposed by many in the United States to be the magic solution for solving all the problems of the world, has turned into a nightmare.

The Importance of Dialectics

That wonderful old philosopher Spinoza — one of the fathers of modern philosophical materialism — once said that the task of philosophy is "neither to weep nor to laugh, but to understand." The Anglo-Saxon world in general has proved remarkably impervious to philosophy. Insofar as they possess any philosophy, the Americans and their English cousins have limited the scope of their thought to the narrow boundaries of empiricism, and its soul-mate pragmatism. Broad generalizations of a more theoretical character were always regarded with something akin to suspicion.

In its day, empiricism played a most progressive (even revolutionary) role in the development of human thought and science. In the sixteenth and seventeenth centuries it marked a decisive break with the stifling intellectual dictatorship of the Church, and laid the basis for the modern scientific method based upon observation and experiment. However, empiricism is helpful only within certain limits. Many people only feel secure when they can refer to the *facts*. Yet, of course, the "facts" do not select themselves! A definite method is required that will help us to look beyond the immediately given and lay bare the processes that lie beyond the "facts."

In the words of that giant Hegel, it is the wish for a rational insight, and not the accumulation of a heap of facts that must possess the mind of one who wishes to adopt the scientific standpoint. The dialectical method provides us with the necessary analytical tools we require to make sense of the mass of information we now possess about nature and society.

Here, however, we are confronted with a difficulty. The most systematic account of dialectics is contained in the writings of Hegel, in particular his massive work *The Science of Logic*. But apart from the highly inaccessible way in which Hegel sets forth his ideas ("abstract and abstruse, Engels called it), the dialectic appears in the hands of Hegel in a mystical, idealist form. It was only rescued by the revolutionary work of Marx and Engels, who for the first time showed the rational kernel in Hegel's thought. In its scientific (materialist) form, the dialectical method provides us with an indispensable tool for understanding the workings of nature, society and human thought.

Marx had always intended to write a work on dialectical materialism, but died before he could do so. After Marx's death, his indefatigable comrade Frederick Engels wrote a number of brilliant studies on dialectical philosophy (*Ludwig Feuerbach and the End of German Classical Philosophy*, *Anti-Dühring*, and *The Dialectics of Nature*). The last-named work was intended to be the basis for a longer work on Marxist philosophy, but unfortunately, Engels was prevented from completing it by the immense work of finishing the second and third volumes of *Capital*, which Marx left unfinished at his death.

Scattered throughout the works of Marx, Engels, Lenin, Trotsky and Plekhanov, one can find a very large amount of material on this subject. But it would take a very long time to extract all this information. The task of putting together a more or less systematic exposition of Marxist philosophy still remains to be done. To the best of my knowledge, *Reason In Revolt* is the first attempt to apply the method of dialectical materialism to the results of modern science since *The Dialectics of Nature*.

In his book *Anti-Dühring*, Engels pointed out that in the last analysis, nature works dialectically. The advances of science over the last hundred years have completely borne out this assertion. And American scientists have been in the forefront of some of the most important areas in modern science. I am thinking in particular of the work of R.C. Lewontin in the field of genetics, and above all the writings of Stephen Jay Gould, news of whose tragic death reached me as I was finishing this introduction.

The latest discoveries of paleontology, in particular the pioneering work Stephen Gould (*punctuated equilibria*) have fundamentally modified the old view of evolution as a slow, gradual process, uninterrupted by sudden catastrophes and leaps. Gould himself was clearly influenced by the ideas of Marxism, and has paid tribute to the contribution of Frederick Engels, who, in his little masterpiece *The Part Played by Labor in the Transition from Ape to Man*, brilliantly anticipated the latest discoveries in the investigation of human origins.

Marxism and Religion

In the first part of *Reason in Revolt*, a reference is made to the contradiction between the marvelous advances of science and the extraordinary lag in human consciousness. This contradiction is particularly striking in the United States. In the country that has done more than any other to advance the cause of science in the past period, the overwhelming majority of people in the USA believe in god, or are religious in some way. This situation is quite different to that of most European countries, where organized religion is dying on its feet (although there is still plenty of superstition and mysticism around).

The degree of scientific and technological advance in the USA is unequalled by any other country. Here we have a tantalizing glimpse of the future — the staggering potential of human development. But we also see a contradiction. Side by side with the most advanced ideas we see the persistence of ideas that have been handed down, unchanged, from a remote and barbarous past.

Since the book first appeared, there have been a number of other spectacular advances in science — notably the mapping of the human genome. These results have completely demolished the positions of genetic determinism that we criticized in *Reason in Revolt*. It has also cut the ground from under the feet of the racist "theories" put forward by certain writers in the USA who attempted to enlist the service of genetics to peddle their reactionary pseudo-scientific "theories," that black people are genetically predisposed to ignorance and poverty. They have also dealt a mortal blow to the nonsense of the Creationists who want to reject Darwinism in favor of the first chapters of *Genesis*, and impose this on American schools.

For many Americans, Marxism is a closed book because it is seen as anti-religious. After all, did Marx not describe religion as the "opium of the people"? As a matter of fact, just before these famous words, Marx wrote: "Religious distress is at the same time the *expression* of real distress and the *protest* against real distress." In essence, religion is an expression of a desire for a better world and a belief that there must be something more to life than the vale of tears through which we pass in the all too brief interval from cradle to grave.

Many people are discontented with their lives. It is not just a question of material poverty — although that exists in the USA as in all other countries. It is also a question of spiritual poverty: the emptiness of people's lives, the mind-deadening routine of work that is just so many hours out of one's life; the alienation that divides men and women from each other; the absence of human relations and solidarity that is deliberately fostered in a society that proudly proclaims the laws of the jungle and the so-called survival of the fittest (read:

wealthiest); the mind-numbing banality of a commercialized "culture." In this kind of world the question we should be asking ourselves is not "is there a life after death" but rather "is there a life *before* death?"

The capitalist system is a monstrously oppressive and inhuman system, which means untold misery, disease, oppression and death for millions of people in the world. It is surely the duty of any humane person to support the fight against such a system. However, in order to fight effectively, it is necessary to work out a serious program, policy and perspective that can guarantee success. We believe that only Marxism (scientific socialism) provides such a perspective.

The problem a Marxist like myself has with religion is basically this: We believe that men and women should fight to transform their lives and to create a genuinely human society which would permit the human race to lift itself up to its true stature. We believe that man has only one life, and should dedicate himself to making this life beautiful and self-fulfilling. If you like, we are fighting for a paradise *on this Earth*, because we do not think there is any other.

Although from a philosophical point of view, Marxism is incompatible with religion, it goes without saying that we are opposed to any idea of prohibiting or repressing religion. We stand for the complete freedom of the individual to hold any religious belief, or none at all. What we do say is that there should be a radical separation between church and state. The churches must not be supported directly or indirectly out of taxation, nor should religion be taught in state schools. If people want religion, they should maintain their churches exclusively through the contributions of the congregation and preach their doctrines on their own time.

To the degree that men and women are able to take control of their lives and develop themselves as free human beings, I believe that interest in religion — that is, the search for consolation in an afterlife — will decline naturally of itself. Of course, you may disagree with this prediction. Time will tell which of us is right. In the meantime, disagreements on such matters should not prevent all honest Christians from joining hands with the Marxists in the struggle for a new and better world.

The Big Bang

There was one part of *Reason in Revolt* that some found rather hard to digest — namely the section on cosmology, where we argued against the theory of the big bang. The standard model of the universe seemed to be so entrenched that it was apparently unassailable. The overwhelming majority accepted it uncritically. To call it into question seemed unthinkable.

The Big Bang theory was an attempt to explain the history of the Universe on the basis of certain observed phenomena, in particular the fact that we can see the galaxies receding from each other. Because of this, most astronomers believe that these star groupings were closer together in the past. If we run the film backwards then all matter, space and time would have erupted from a point in a massive explosion, involving staggering amounts of energy.

In the most widely accepted cosmological model, called the inflationary model, the universe was said to have been born in an instantaneous creation of matter and energy. It is the modern equivalent of the old religious dogma of the creation of the world from nothing. *The Big Bang is alleged to be the beginning of space, matter and time.* As the universe has inflated since that event, matter and energy have spread out in clumps. The spreading could potentially continue forever.

In fact, there are serious problems with this theory, which we outlined in detail in *Reason in Revolt*. In particular, questions about what happened *before* the Big Bang cannot really be asked because there is supposed to have been no "before" — since there was no time. In this way, an absolute limitation is placed on the possibility of our understanding the Universe, thus leaving the door wide open for all kinds of mystical ideas — which have been pouring out in vast quantities in recent years. Nevertheless, the inflationary theory has survived since it was introduced in the late 1970s, while cosmologists have discarded competing ideas one by one.

However, new problems with the existing theory are becoming apparent all the time. The latest was in 1998, when studies of distant, exploding stars showed the Universe was *expanding at an accelerating rate.* This was a big surprise, since most researchers believed that either the Universe would expand forever at the same rate or else slow down and contract, eventually coming back together in a "Big Crunch."

A recent report by Paul J. Steinhardt and his colleague Neil Turok of Cambridge University posted on April 25th, 2002, on the website of the prestigious journal *Science*, has thrown down a serious challenge to the accepted wisdom. The two scientists have put forward a new model to explain how the cosmos is and where it might be going. They argue — as we did in *Reason in Revolt* — that *the Universe had no beginning and it will have no end.* Steinhardt and Turok, point out that the standard model has several shortcomings. It cannot tell us what happened before the Big Bang or explain the eventual fate of the Universe. Will it expand forever or stop and contract? These were some of the objections we raised in *Reason in Revolt*.

The new model offers a streamlined alternative to the standard model. It treats the Big Bang not as the moment of creation, but as a transition between two cycles in an endless process of cosmological rebirth. According to the model, the Big Bang is followed by a period of slow expansion and gradual accumulation of dark energy. As dark energy becomes dominant, it stimulates cosmic acceleration. The authors maintain that the current era is near the transition between these stages.

At present, they argue, the Universe is in an expansionary phase, and the current expansion will go on for trillions of years, before reaching a critical point where the process takes a new direction. Although there are many questions still to be answered (in particular the question of this hypothetical "dark energy"), the new model seems to be a vast improvement on the existing one, which states that the big bang was the beginning of time, matter, space and energy — clearly a mystical and unscientific conception. The new theory does away with the idea that the Universe has either a beginning or an end — it is infinite in both time and space. This model therefore puts an end to the nonsense of the creation of the Universe from nothing:

"What we're proposing in this new picture is that the Big Bang is not a beginning of time but really just the latest in an infinite series of cycles, in which the Universe has gone through periods of heating, expanding, cooling, stagnating, emptying, and then re-expanding again" (BBC report).

The picture of the Universe presented here is one that is entirely consistent with the theories of dialectical materialism, which state that the Universe is *infinite, eternal, and ever-changing*. This does not at all preclude the possibility of a big bang. Indeed, we have already argued that there have probably been many big bangs. But what it certainly does preclude is any question of matter (or energy, which is exactly the same thing) being created out of nothing (as the Big Bang implies) or destroyed.

It is too early to say whether it will be verified in detail. However, what is clear is that the deficiencies of the Big Bang theory are now becoming clear, and the search is on for an alternative. Whether or not the present theory is correct in its detail, the method that its authors have used — a materialist and dialectical method — is obviously correct. And, as they correctly write in *Science*: "*The ultimate arbiter will be Nature.*"

Marxism and the Future

Marxism is a philosophy, but it is quite unlike other philosophies. Dialectical materialism is both a powerful methodological tool to understand the workings of nature, thought and human society and a guide to action. As the young Marx put it: "philosophers have only interpreted the world in various ways — *the point is, however, to change it.*"

Now, it may be that you are quite happy with the world in which we live, and do not wish to change it. In that case, you may find the present work educational, or at least entertaining. But you will not have understood it, basically because we will be talking mutually unintelligible languages. However, if ever there was a time when Americans should be seriously re-examining their view of the world and their place in it, that time is now. And in order to obtain a rational insight into this world a knowledge of dialectical materialism is of great importance.

The most essential feature of dialectical materialism is its dynamic character. It sees the world as an ever-changing process, driven by internal contradictions, in which sooner or later things change into their opposite. Moreover, the line of development is not a smooth, linear process, but a line that is periodically interrupted by sudden leaps, explosions that transform quantity into quality. This is an accurate picture of both processes we see in nature and in the process of social development we call history.

Most people imagine that the kind of world into which they are born is something fixed and immutable. They rarely question its values, its morality, its religion, its political and state institutions. This mental inertia, reinforced by the dead weight of tradition, customs, habit and routine, is a powerful cement that permits a given socio-economic order to continue to exist long after it has lost its rational basis. In the USA, perhaps more than any other country in the world, this inertia exercises a major role and prevents people from realizing what is happening to them.

In actual fact, societies are not immutable. The whole of history teaches us that. Socio-economic systems, like individual men and women, are born, mature, reach a high point in their development, and then at a certain point enter into a phase of decline and decay. When a society ceases to play a progressive role (which, in the last analysis, is that point where it is unable to develop the productive forces as it did in the past), people can feel it. It manifests itself in all manner of ways — not only in the economic field. The old morality begins to break down. There is a crisis of the family and personal relations, a growing lack of solidarity and social cohesion, a rise of crime and violence. People no longer believe in the old religions and turn in the direction of mysticism, superstition and exotic sects. We have seen these things many times in history, and we are seeing the same things now — even in the USA.

We are living at a time when many people have begun to ask questions about the world in which they live, and to ask questions is never a bad thing. The terrible events of September 11, 2001 have caused many Americans to think seriously about matters in which they previously showed little interest. They have suddenly realized that all is not well with the world, and that America is deeply involved in a worldwide crisis from which no one can escape, and in which no one is safe. The destruction of the twin towers cast a dark shadow over

America. For a time, Bush and the most reactionary wing of the ruling class have had things all their own way. But this situation will not last forever. Sooner or later the thick fog of propaganda and lies will dissipate and people will become aware of the real state of affairs both in the USA and on a world scale.

Although many people feel in their innermost being that something is going badly wrong, they find no logical explanation for it. That is not surprising. The entire way in which they have been taught to think from their earliest years conditions them to reject any suggestion that there is something fundamentally wrong with the society in which they live. They will close their eyes and try to avoid drawing uncomfortable conclusions for as long as they can.

This is quite natural. Is very hard for people to question the beliefs they have been brought up with. But sooner or later, events catch up with them — cataclysmic events that compel them to re-think many things that they previously took for granted. And when such a moment arrives, the same people who stubbornly refused to consider new ideas, will eagerly examine what only yesterday they regarded as heresies, and find in them the explanations and alternatives for which they were striving.

Today, Marxism is seen as such a heresy. Every hand is raised against it. It is said to have no basis, to have failed, to be out of date. But if this is really the case, then why do the apologists of capitalism still persist in attacking it? Surely, if it is so dead and irrelevant, they should just ignore it. The power of Marxist ideas is precisely that they – and they alone — can provide a coherent, rigorous, and, yes, scientific explanation of the most important phenomena of the world in which we live.

It is a matter of great regret that so many people, especially in the USA, have the same attitude towards Marxism as the representatives of the Roman Catholic Church had towards Galileo's telescope. When Galileo begged them to look with their own eyes and examine the evidence, they stubbornly refused to do so. They just knew that Galileo was wrong, and that was that. In the same way, many people "just know" that Marxism is wrong, and do not see any reason to investigate the matter any further. But if Marxism is wrong, by studying it, you will be more firmly convinced of its erroneousness. You have nothing to lose, and will have added to your store of knowledge. But the author of these lines is firmly convinced that if more people just took the trouble to read the works of Marx, Engels, Lenin and Trotsky, they would soon convince themselves that Marxism really does have a lot of important things to say — and that these things are of great relevance to the modern world.

In recommending the present work to the American public, it is my fervent hope to convince the reader of the correctness and relevance of the ideas of Marx and Engels in the world of the 21st century. If I succeed even partly in convincing you, I will be very pleased. If not, I hope to have dispelled many misconceptions about Marxism and show that it at least has some interesting things to say about the world in which we live. In any event, I hope it will make people think more critically about our society, its present and its future.

London,
June 6, 2002

Postscript:

Those who wish to continue their studies of Marxism, and more importantly join in the historic struggle for socialism, can visit the In Defence of Marxism website at www.marxist.com and the Socialist Appeal magazine of the Workers International League in the US at www.socialistappeal.org. These sites provide regular Marxist analysis of current events as well as historical and economic analysis, theory, and more. The Youth for International Socialism website at www.newyouth.com also provides a wealth of information and learning material for those wishing to learn more about Marxism and the struggle for world socialism.

Authors' Foreword

"A specter is haunting Europe" (*The Communist Manifesto*)

Mark Twain once joked that rumors of his death had been exaggerated. It is a striking fact that, every year for approximately the last 150 years, Marxism has been pronounced defunct. Yet, for some unaccountable reason, it maintains a stubborn vitality, the best proof of which is the fact that the attacks upon it not only continue, but actually tend to multiply both in frequency and acrimony. If Marxism is really irrelevant, why bother even to mention it? The fact is that the detractors of Marxism are still haunted by the same old specter. They are uncomfortably aware that the system they defend is in serious difficulties, riven by insurmountable contradictions; that the collapse of a totalitarian caricature of socialism is not the end of the story.

In the last few years, ever since the fall of the Berlin Wall, there has been an unprecedented ideological counter-offensive against Marxism, and the idea of socialism in general. Francis Fukuyama went so far as to proclaim the *"End of History."* But history continues, and with a vengeance. The monstrous regime of Stalinism in Russia has been replaced by an even greater monstrosity. The real meaning of *"free market reform"* in the former Soviet Union has been a frightful collapse of the productive forces, science and culture, on a scale which can only be likened to a catastrophic defeat in war.

Despite all this — or maybe because of it — the admirers of the alleged virtues of capitalism are dedicating considerable resources to affirm that the collapse of Stalinism proves that socialism does not work. It is alleged that the entire body of ideas worked out by Marx and Engels, and later developed by Lenin, Trotsky and Rosa Luxemburg, have been completely discredited. Upon closer examination, however, what is becoming increasingly obvious is the crisis of the so-called free-market economy, which currently condemns 22 million human beings to a life of enforced inactivity in the industrialized nations alone, wasting the creative potential of a whole generation. The whole of Western society finds itself in a blind-alley, not only economically, politically and socially, but morally and culturally. The fall of Stalinism, which was predicted by Marxists decades ago, cannot disguise the fact that, in the final decade of the 20th century, the capitalist system is in a deep crisis on a world scale. The strategists of Capital look to the future with profound foreboding. And at bottom the more honest among them ask themselves the question they dare not answer: Was old Karl right after all?

Whether one accepts or rejects the ideas of Marxism, it is impossible to deny the colossal impact which they have exercised on the world. From the appearance of *The Communist Manifesto*, down to the present day, Marxism has been a decisive factor, not only in the political arena, but in the development of human thought. Those who fought against it were nevertheless compelled to take it as their starting point. And, irrespective of the present state of affairs, it is an indisputable fact that the October Revolution changed the entire course of world history. A close acquaintance with the theories of Marxism is therefore a necessary precondition for anyone who wishes to understand some of the most fundamental phenomena of our times.

Engels' Role

August 1995 marks the centenary of the death of Frederick Engels, the man who, together with Karl Marx developed an entirely new way of looking at the world of nature, society and human development. The role played by Engels in the development of Marxist thought is a subject which has never been given its due. This is partly the result of the towering genius of Marx, which inevitably overshadows the contribution made by his lifelong friend and comrade. In part it flows from the innate humility of Engels, who always played down his own contribution, preferring to emphasize Marx's preeminence. At his death, Engels gave instructions that his body be cremated and his ashes cast into the sea at Beachy Head, because he wanted no monument. Like Marx, he heartily detested anything remotely resembling a cult of the personality. The only real monument they wished to leave behind was the imposing body of ideas, which provides a comprehensive ideological basis for the fight for the socialist transformation of society.

Many people do not realize that the scope of Marxism extends far beyond politics and economics. At the heart of Marxism lies the philosophy of dialectical materialism. Unfortunately, the immense labor of writing *Capital* prevented Marx from writing a comprehensive work on the subject, as he had intended. If we exclude the early works, such as *The Holy Family* and *The German Ideology*, which represent important, but still preparatory, attempts to develop a new philosophy, and the three volumes of *Capital*, which are a classic example of the concrete application of the dialectical method to the particular sphere of economics, then the principal works of Marxist philosophy were all the work of Engels. Whoever wants to understand dialectical materialism must begin by a thorough knowledge of *Anti-Dühring*, *The Dialectics of Nature*, and *Ludwig Feuerbach*.

To what extent have the philosophical writings of this man who died a century ago stood the test of time? That is the starting point of the present work. Engels defined dialectics as *"the most general laws of motion of nature, society, and human thought."* In *The Dialectics of Nature*, in particular, Engels based himself on a careful study of the most advanced scientific knowledge of the day, to show that *"in the last analysis, the workings of nature are dialectical."* It is the contention of the present work that the most important discoveries of 20th century science provides a striking confirmation of this.

What is most amazing is not the attacks on Marxism, but the complete ignorance of it which is displayed by its detractors. Whereas no one would dream of practicing as a car mechanic without studying mechanics, everyone feels free to express an opinion about Marxism, without any knowledge of it whatsoever. The present work is an attempt to explain the basic ideas of Marxist philosophy, and show the relation between it and the position of science and philosophy in the modern world. The intention of the authors is to produce a trilogy, which will cover the three main component parts of Marxism: 1) Marxist philosophy (dialectical materialism), 2) the Marxist theory of history and society (historical materialism), and 3) Marxist economics (the labor theory of value).

Originally, we intended to include a section on the history of philosophy, but in view of the length of the present work we have decided to publish this separately. We begin with a review of the philosophy of Marxism, dialectical materialism. This is fundamental because it is the method of Marxism. Historical materialism is the application of this method to the study of the development of human society; the labor theory of value is the result of the application of the same method to economics. An understanding of Marxism is impossible without a grasp of dialectical materialism.

The ultimate proof of dialectics is nature itself. The study of science occupied the attention of Marx and Engels all their lives. Engels had intended to produce a major work, outlining in detail the relation between dialectical materialism and science, but was prevented from completing it because of the heavy burden of work on the second and third volumes of *Capital*, left unfinished when Marx died. His incomplete manuscripts for *The Dialectics of Nature* were only published in 1925. Even in their unfinished state, they provide a most important source for the study of Marxist philosophy, and provide brilliant insights into the central problems of science.

One of the problems we faced in writing the present work is the fact that most people have only a second-hand knowledge of the basic writings of Marxism. This is regrettable, since the only way to understand Marxism is by reading the works of Marx, Engels, Lenin and Trotsky. The great majority of works that purport to explain *"what Marx meant"* are worthless. We have therefore decided to include a large number of quite lengthy quotes, particularly from Engels, partly to give the reader direct access to these ideas without any "translation," and partly in the hope that it will stimulate people to read the originals for themselves. This method does not make the book easier to read, but was, in our opinion, necessary. In the same way, we felt obliged to reproduce some lengthy quotes of authors with whom we disagree, on the principle that it is always better to allow one's opponents to speak for themselves.

London,
May 1, 1995

Part One: Reason and Unreason

1. Introduction

We are living in a period of profound historical change. After a period of 40 years of unprecedented economic growth, the market economy is reaching its limits. At the dawn of capitalism, despite its barbarous crimes, it revolutionized the productive forces, thus laying the basis for a new system of society. The First World War and the Russian Revolution signaled a decisive change in the historical role of capitalism. From a means of developing the productive forces, it became transformed into a gigantic fetter upon economic and social development. The period of upswing in the West in the period of 1948-73 seemed to promise a new dawn. Even so, the benefits were limited to a handful of developed capitalist countries. For two-thirds of humanity living in the Third World, the picture was one of mass unemployment, poverty, wars and exploitation on an unprecedented scale. This period of capitalism ended with the so-called "oil crisis" of 1973-4. Since then, they have not managed to get back to the kind of growth and levels of employment they had achieved in the post-war period.

A social system in a state of irreversible decline expresses itself in cultural decay. This is reflected in a hundred different ways. A general mood of anxiety and pessimism as regards the future spreads, especially among the intelligentsia. Those who yesterday talked confidently about the inevitability of human progress and evolution, now see only darkness and uncertainty. The 20th century is staggering to a close, having witnessed two terrible world wars, economic collapse and the nightmare of fascism in the period between the wars. These were already a stern warning that the progressive phase of capitalism was past.

The crisis of capitalism pervades all levels of life. It is not merely an economic phenomenon. It is reflected in speculation and corruption, drug abuse, violence, all-pervasive egotism and indifference to the suffering of others, the breakdown of the bourgeois family, the crisis of bourgeois morality, culture and philosophy. How could it be otherwise? One of the symptoms of a social system in crisis is that the ruling class increasingly feels itself to be a fetter on the development of society.

Marx pointed out that the ruling ideas of any society are the ideas of the ruling class. In its heyday, the bourgeoisie not only played a progressive role in pushing back the frontiers of civilization, but was well aware of the fact. Now the strategists of capital are seized with pessimism. They are the representatives of an historically doomed system, but cannot reconcile themselves to the fact. This central contradiction is the decisive factor which sets its imprint upon the mode of thinking of the bourgeoisie today. Lenin once said that a man on the edge of a cliff does not reason.

Lag in Consciousness

Contrary to the prejudice of philosophical idealism, human consciousness in general is extraordinarily conservative, and always tends to lag far behind the development of society, technology and the productive forces. Habit, routine, and tradition, to use a phrase of Marx, weigh like an Alp on the minds of men and women, who, in "normal" historical periods cling stubbornly to the well-trodden paths, from an instinct of self-preservation, the roots of which lie in the remote past of the species. Only in exceptional periods of history, when the social and moral order begin to crack under the strain of intolerable pressures do the mass of people start to question the world into which they have been born, and to doubt the beliefs and prejudices of a lifetime.

Such a period was the epoch of the birth of capitalism, heralded by the great cultural re-awakening and spiritual regeneration of Europe after its lengthy winter sleep under feudalism. In the period of its historical ascent, the bourgeoisie played a most progressive role, not only in developing the productive forces, and thereby mightily expanding humanity's power over nature, but also in extending the frontiers of science, knowledge and culture. Luther, Michelangelo, Leonardo, Dührer, Bacon, Kepler, Galileo and a host of other pathfinders of civilization shine like a galaxy illuminating the broad highroad of human cultural and scientific advance opened by the Reformation and Renaissance. However, such revolutionary periods do not come into being easily or automatically. The price of progress is struggle — the struggle of the new against the old, the living against the dead, the future against the past.

The rise of the bourgeoisie in Italy, Holland, England and later in France was accompanied by an extraordinary flourishing of culture, art and science. One would have to look back to ancient Athens to find a precedent for this. Particularly in those countries where the bourgeois revolution triumphed in the 17th and 18th centuries, the development of the forces of production and technology was accompanied by a parallel development of science and thought, which drastically undermined the ideological domination of the Church.

In France, the classical country of the bourgeois revolution in its political expression, the bourgeoisie in 1789-93 carried out its revolution under the banner of Reason. Long before it toppled the formidable walls of the Bastille, it was necessary to overthrow the invisible but no less formidable walls of religious superstition in the minds of men and women. In its revolutionary youth the French bourgeoisie was rationalist and atheist. Only after installing themselves in power did the men of property, finding themselves confronted by a new revolutionary class, jettison the ideological baggage of their youth.

Not long ago France celebrated the two hundredth anniversary of its great revolution. It was curious to note how even the memory of a revolution two centuries ago fills the establishment with unease. The attitude of the French ruling class to their own revolution vividly recalled that of an old libertine who tries to gain a ticket to respectability — and perhaps admittance to heaven — by renouncing the sins of his youth which he is no longer in a position to repeat. Like all established privileged classes, the capitalist class seeks to justify its existence, not only to society at large, but to itself. In its search for ideological points of support, which would tend to justify the status quo and sanctify existing social relations, they rapidly rediscovered the enchantments of Mother Church, particularly after the mortal terror they experienced at the time of the Paris Commune. The church of Sacré Coeur is a concrete expression of the bourgeois' fear of revolution translated into the language of architectural philistinism.

Marx (1818-83) and Engels (1820-95) explained that the fundamental driving force of all human progress is the development of the productive forces - industry, agriculture, science and technique. This is a truly great theoretical generalization without which it is impossible to understand the movement of human history in general. However, it does not mean, as dishonest or ignorant detractors of Marxism have attempted to show, that Marx *"reduces everything to economics."* Dialectical and historical materialism takes full account of phenomena such as religion, art, science, morality, law, politics, tradition, national characteristics and all the other manifold manifestations of human consciousness. But not only that. It shows their real content and how they relate to the actual development of society, which in the last analysis clearly depends upon its capacity to reproduce and expand the material conditions for its existence. On this subject, Engels wrote the following:

> According to the materialist conception of history, the ultimately determining element in history is the production and reproduction of real life. More than this neither Marx nor I have ever asserted. Hence, if someone twists this into saying that the economic element is the only determining one, he transforms that position into a meaningless, abstract, senseless phrase. The economic situation is the basis, but the various elements of the superstructure — political forms of the class struggle and its results, to wit: constitutions established by victorious classes after a successful battle, etc., judicial forms, and the reflexes of all these actual struggles in the brains of the participants, political, juristic, philosophical theories, religious views and their further development into systems of dogmas also exercise their influence upon the course of the historical struggles, and in many cases predominate in determining their form. [1]

The affirmation of historical materialism that, in general, human consciousness tends to lag behind the development of the productive forces seems paradoxical to some. Yet it is graphically expressed in all kinds of ways in the United States where the achievements of science have reached their highest level. The constant advance of technology is the prior condition for bringing about the real emancipation of men and women, through the establishment of a rational socioeconomic system, in which human beings exercise conscious control over their lives and environment. Here, however, the contrast between the rapid development of science and technology and the extraordinary lag in human thinking presents itself in its most glaring form.

In the USA nine persons out of ten believe in the existence of a supreme being, and seven out of ten in a life after death. When the first American astronaut who succeeded in circumnavigating the world in a spacecraft was asked to broadcast a message to the inhabitants of the earth, he made a significant choice. Out of the whole of world literature, he chose the first sentence of the book of *Genesis*: "In the beginning, God created heaven and the earth." This man, sitting in his space ship, a product of the most advanced technology ever seen, had his mind full to the brim with superstitions and phantoms handed down with little change from the primeval past.

Seventy years ago, in the notorious "monkey trial" of 1925, a teacher called John Scopes was found guilty of teaching the theory of evolution, in contravention of the laws of the state of Tennessee. The trial actually upheld the state's anti-evolution laws, which were not abolished until 1968, when the US Supreme Court ruled that the teaching of creation theories was a violation of the constitutional ban on the teaching of religion in state schools. Since then, the creationists changed their tactics, trying to turn creationism into a *"science."* In this, they have the support, not only of a wide layer of public opinion, but of not a few scientists, who are prepared to place their services at the disposal of religion in its most crude and obscurantist form.

In 1981 American scientists, making use of Kepler's laws of planetary motion, launched a spacecraft that made a spectacular rendezvous with Saturn. In the same year an American judge had to declare unconstitutional a law passed in the state of Arkansas which imposed on schools the obligation to treat so-called "creation-science" on equal terms with the theory of evolution. Among other things, the creationists demanded the recognition of Noah's flood as a primary geological agent. In the course of the trial, witnesses for the defense expressed fervent belief in Satan and the possibility that life was brought to earth in meteorites, the variety of species being explained by a kind of meteoric shuttle-service! At the trial, Mr. N. K. Wickremasinge of the University of Wales was quoted as saying that insects might be more intelligent than humans, although "they're not letting on...because things are going so well for them." [2]

The religious fundamentalist lobby in the USA has mass support, access to unlimited funds, and the backing of congressmen. Evangelical crooks make fortunes out of radio stations with a following of millions. The fact that in the last decade of the 20th century there are a large number of educated men and women — including scientists -in the most technologically advanced country the world has ever known who are prepared to fight for the idea that the book of *Genesis* is literally true, that the universe was created in six days about 6,000 years ago, is, in itself, a most remarkable example of the workings of the dialectic.

"Reason Becomes Unreason"

The period when the capitalist class stood for a rational world outlook has become a dim memory. In the epoch of the senile decay of capitalism, the earlier processes have been thrown into reverse. In the words of Hegel, *"Reason becomes Unreason."* It is true that, in the industrialized countries, "official" religion is dying on its feet. The churches are empty and increasingly in crisis. Instead, we see a veritable "Egyptian plague" of peculiar religious sects, accompanied by the flourishing of mysticism and all kinds of superstition. The frightful epidemic of religious fundamentalism — Christian, Jewish, Islamic, Hindu — is a graphic manifestation of the impasse of society. As the new century beckons, we observe the most horrific throwbacks to the Dark Ages.

This phenomenon is not confined to Iran, India and Algeria. In the United States we saw the "Waco massacre," and after that, in Switzerland, the collective suicide of another group of religious fanatics. In other Western countries, we see the uncontrolled spread of religious sects, superstition, astrology and all kinds of irrational tendencies. In France, there are about 36,000 Catholic priests, and over 40,000 professional astrologers who declared their earnings to the taxman. Until recently, Japan appeared to be an exception to the rule. William Rees-Mogg, former editor of the London Times, and arch-Conservative, in his recent book *The Great Reckoning, How the World Will Change in the Depression of the 1990s* states that: *"The revival of religion is something that is happening throughout the world in varying degrees. Japan may be an exception, perhaps because social order has as yet shown no signs of breaking down there..."* [3] Rees-Mogg spoke too soon. A couple of years after these lines were written, the horrific gas attack on the Tokyo underground drew the world's attention to the existence of sizable groups of religious fanatics even in Japan, where the economic crisis has put an end to the long period of full employment and social stability. All these phenomena bear a striking resemblance to what occurred in the period of the decline of the Roman Empire. Let no one object that such things are confined to the fringes of society. Ronald and Nancy Reagan regularly consulted astrologers about all their actions, big and small. Here are a couple of extracts from Donald Regan's book, *For the Record*:

> Virtually every major move and decision the Reagans made during my time as White House chief of staff was cleared in advance with a woman in San Francisco who drew up horoscopes to make certain that the planets were in a favorable alignment for the enterprise. Nancy Reagan seemed to have absolute faith in the clairvoyant powers of this woman, who had predicted that 'something' bad was going to happen to the president shortly before he was wounded in an assassination attempt in 1981.

> Although I had never met this seer — Mrs. Reagan passed along her prognostications to me after conferring with her on the telephone — she had become such a factor in my work, and in the highest affairs of the state at one point I kept a color-coded calendar on my desk (numerals highlighted in green ink for 'good' days, red for 'bad' days, yellow for 'iffy' days) as an aid to remember when it was propitious to move the president of the United States from one place to another, or schedule him to speak in public, or commence negotiations with a foreign power.

> Before I came to the White House, Mike Deaver had been the man who integrated the horoscopes of Mrs. Reagan's into the presidential schedule...It is a measure of his discretion and loyalty that few in the White House knew that Mrs. Reagan was even part of the problem [waiting for schedules] — much less that an astrologer in San Francisco was approving the details of the presidential schedule. Deaver told me that Mrs. Reagan's dependence on the occult went back at least as far as her husband's governorship, when she had relied on the advice of the famous Jeane Dixon. Subsequently, she had lost confidence in Dixon's powers. But the First Lady seemed to have absolute faith in the clairvoyant talents of the woman in San Francisco. Apparently, Deaver had ceased to think there was anything remarkable about

this long-established floating séance...To him it was simply one of the little problems in the life of a servant of the great. "At least," he said, "this astrologer is not as kooky as the last one."

Astrology was used in the planning of the summit between Reagan and Gorbachev, according to the family soothsayer, but things didn't go smoothly between the two first ladies because Raisa's birth date was unknown! The movement in the direction of a "free market economy" in Russia has since bestowed the blessings of capitalist civilization on that unfortunate country — mass unemployment, social disintegration, prostitution, the mafia, an unprecedented crime wave, drugs and religion. It has recently emerged that Yeltsin himself consults astrologers. In this respect also, the nascent capitalist class in Russia has shown itself to be an apt pupil of its Western role models.

The prevailing sense of disorientation and pessimism finds its reflection in all sorts of ways, not only directly in politics. This all-pervasive irrationality is not an accident. It is the psychological reflection of a world where the destiny of humanity is controlled by terrifying and seemingly invisible forces. Just look at the sudden panic on the stock exchange, with *"respectable"* men and women scurrying around like ants when their nest is broken open. These periodic spasms causing a herd-like panic are a graphic illustration of capitalist anarchy. And this is what determines the lives of millions of people. We live in the midst of a society in decline. The evidence of decay is present on all sides. Conservative reactionaries bemoan the breakdown of the family and the epidemic of drugs, crime, mindless violence, and the rest. Their only answer is to step up state repression — more police, more prisons, harsher punishments, even genetic investigation of alleged "criminal types." What they cannot or will not see is that these phenomena are the symptoms of the blind alley of the social system which they represent.

These are the defenders of *"market forces,"* the same irrational forces that presently condemn millions of people to unemployment. They are the prophets of "supply-side" economics, which John Galbraith shrewdly defined as the theory that the poor have too much money, and the rich too little. The prevailing "morality" is that of the market place, that is, the morality of the jungle. The wealth of society is concentrated into fewer and fewer hands, despite all the demagogic nonsense about a "property-owning democracy" and "small is beautiful." We are supposed to live in a democracy. Yet a handful of big banks, monopolies, and stock exchange speculators (generally the same people) decide the fate of millions. This tiny minority possesses powerful means of manipulating public opinion. They have a monopoly of the means of communication, the press, radio and television. Then there is the spiritual police — the church, which for generations has taught people to look for salvation in another world.

Science and the Crisis of Society

Until quite recently, it appeared that the world of science stood aloof from the general decay of capitalism. The marvels of modern technology conferred colossal prestige upon scientists, who appeared to be endowed with almost magical qualities. The respect enjoyed by the scientific community increased in the same proportion as their theories became increasingly incomprehensible to the majority of even educated people. However, scientists are ordinary mortals who live in the same world as the rest of us. As such, they can be influenced by prevailing ideas, philosophies, politics and prejudices, not to speak of sometimes very substantial material interests.

For a long time it was tacitly assumed that scientists — especially theoretical physicists — were a special sort of people, standing above the common run of humanity, and privy to the mysteries of the universe denied to ordinary mortals. This 20th century myth is well conveyed by the old science-fiction movies, where the earth was always threatened with annihilation by aliens from outer space (in reality, the threat to the future of humankind comes from a source much nearer to home, but that is another story). At the last moment, a man in a white coat always turns up, writes a complicated equation on the blackboard, and the problem is fixed in no time at all.

The truth is rather different. Scientists and other intellectuals are not immune to the general tendencies at work in society. The fact that most of them profess indifference to politics and philosophy only means that they fall prey more easily to the current prejudices which surround them. All too often their ideas can be used to support the most reactionary political positions. This is particularly clear in the field of genetics where a veritable counter-revolution has taken place, particularly in the United States. Allegedly scientific theories are being used to *"prove"* that criminality is caused, not by social conditions, but by a "criminal gene." Black people are alleged to be disadvantaged, not because of discrimination, but because of their genetic make-up. Similar arguments are used for poor people, single mothers, women, homosexuals, and so on. Of course, such *"science"* is highly convenient to the Republican-dominated Congress intent on ruthlessly cutting welfare.

The present book is about philosophy — more precisely, the philosophy of Marxism, dialectical materialism. It is not the business of philosophy to tell scientists what to think and write, at least when they write about science. But scientists have a habit of expressing opinions about all kinds of things — philosophy, religion, politics. This they are perfectly entitled to do. But when they use what may well be perfectly sound scientific credentials in order to defend extremely unsound and reactionary philosophical views, it is time to put things in their context. These pronouncements do not remain among a handful of professors. They are seized upon by right wing politicians, racists and religious fanatics, who attempt to cover their backsides with pseudo-scientific arguments.

Scientists frequently complain that they are misunderstood. They do not mean to provide ammunition for mystical charlatans and political crooks. That may be so. But in that case, they are guilty of culpable negligence or, at the very least, astounding naïvety. On the other hand, those who make use of the erroneous philosophical views of scientists cannot be accused of naïvety. They know just where they stand. Rees-Mogg argues that *"as the religion of secular consumerism is left behind like a rusting tail fin, sterner religions that involve real moral principles and angry gods will make a comeback. For the first time in centuries, the revelations of science will seem to enhance rather than undermine the spiritual dimension in life."* For Rees-Mogg religion is a useful weapon to keep the underprivileged in their place, alongside the police and prison service. He is commendably blunt about it:

> The lower the prospect of upward mobility, the more rational it is for the poor to adopt an anti-scientific, delusional world view. In place of technology, they employ magic. In place of independent investigation, they opt for orthodoxy. Instead of history, they prefer myths. In place of biography, they venerate heroes. And they generally substitute kin-based behavioral allegiances for the impersonal honesty required by the market. [4]

Let us leave aside the unconsciously humorous remark about the *"impersonal honesty"* of the market place, and concentrate on the core of his argument. At least Rees-Mogg does not try to conceal his real intentions or his class standpoint. Here we have the utmost frankness from a defender of the establishment. The creation of an under-class of poor, unemployed, mainly black people, living in slums, presents a potentially explosive threat to the existing social order. The poor, fortunately for us, are ignorant. They must be kept in ignorance, and encouraged in their superstitious and religious delusions which we of the "educated classes" naturally do not share! The message, of course, is not new. The same song has been sung by the rich and powerful for centuries. But what is significant is the reference to science, which, as Rees-Mogg indicates, is now regarded for the first time as an important ally of religion.

Recently, theoretical physicist Paul Davies was awarded £650,000 by the Templeton Prize for Progress in Religion, for showing *"extraordinary originality"* in advancing humankind's understanding of God or spirituality. Previous winners include Alexander Solzhenitsyn, Mother Teresa, evangelist Billy Graham, and the Watergate burglar-turned-preacher Charles Colson. Davies, author of such books as *God and the New Physics*, *The Mind of God*, and *The Last Three Minutes*, insists that he is "not a religious person in the conventional sense" (whatever that might mean), but he maintains that *"science offers a surer path to God than religion."* [5]

Despite Davies' ifs and buts, it is clear that he represents a definite trend, which is attempting to inject mysticism and religion into science. This is not an isolated phenomenon. It is becoming all too common, especially in the field of theoretical physics and cosmology, both heavily dependent upon abstract mathematical models which are increasingly seen as a substitute for empirical investigation of the real world. For every conscious peddler of mysticism in this field, there are a hundred conscientious scientists, who would be horrified to be identified with such obscurantism. The only real defense against idealist mysticism, however, is a consciously materialist philosophy — the philosophy of dialectical materialism.

It is the intention of this book to explain the basic ideas of dialectical materialism, first worked out by Marx and Engels, and show their relevance to the modern world, and to science in particular. We do not pretend to be neutral. Just as Rees-Mogg defends the interests of the class he represents, and makes no bones about it, so we openly declare ourselves as the opponents of the so-called *"market economy"* and all that it stands for. We are active participants in the fight to change society. But before we can change the world, one has to understand it. It is necessary to conduct an implacable struggle against all attempts to confuse the minds of men and women with mystical beliefs which have their origin in the murky prehistory of human thought. Science grew and developed to the degree that it turned its back on the accumulated prejudices of the past. We must stand firm against this attempt to put the clock back four hundred years.

A growing number of scientists are becoming dissatisfied with the present situation, not only in science and education, but in society at large. They see the contradiction between the colossal potential of technology and a world where millions of people live on the border line of starvation. They see the systematic misuse of science in the interest of profit for the big monopolies. And they must be profoundly disturbed by the continuous attempts to dragoon the scientists into the service of religious obscurantism and reactionary social policies. Many of them were repelled by the bureaucratic and totalitarian nature of Stalinism. But the collapse of the Soviet Union has shown that the capitalist alternative is even worse. By their own experience, many scientists will come to the conclusion that the only way out of the social, economic, and cultural impasse is by means of some kind of rational planned society, in which science and technology is put at the disposal of humanity, not private profit. Such a society must be democratic, in the real sense of the word, involving the conscious control and participation of the entire population. Socialism is democratic by its very nature. As Trotsky pointed out *"a nationalized planned economy needs democracy, as the human body needs oxygen."*

It is not enough to contemplate the problems of the world. It is necessary to change it. First, however, it is necessary to understand the reason why things are as they are. Only the body of ideas worked out by Marx and Engels, and subsequently developed by Lenin and Trotsky can provide us with the adequate means of achieving this understanding. We believe that the most conscious members of the scientific community, through their own work and experience, will come to realize the need for a consistently materialist world outlook. That is offered by dialectical materialism. The recent advances of the theories of chaos and complexity show that an increasing number of scientists are moving in the direction of dialectical thinking. This is an enormously significant development. There is no doubt that new discoveries will deepen and strengthen this trend. We are firmly convinced that dialectical materialism is the philosophy of the future.

2.   PHILOSOPHY AND RELIGION

Do We Need Philosophy?

Before we start, you may be tempted to ask, "Well, what of it?" Is it really necessary for us to bother about complicated questions of science and philosophy? To such a question, two replies are possible. If what is meant is: do we need to know about such things in order to go about our daily life, then the answer is evidently no. But if we wish to gain a rational understanding of the world in which we live, and the fundamental processes at work in nature, society and our own way of thinking, then matters appear in quite a different light.

Strangely enough, everyone has a "philosophy." A philosophy is a way of looking at the world. We all believe we know how to distinguish right from wrong, good from bad. These are, however, very complicated issues which have occupied the attention of the greatest minds in history. When confronted with the terrible fact of the existence of events like the fratricidal war in the former Yugoslavia, the re-emergence of mass unemployment, the slaughter in Rwanda, many people will confess that they do not comprehend such things, and will frequently resort to vague references to "human nature." But what is this mysterious human nature which is seen as the source of all our ills and is alleged to be eternally unchangeable? This is a profoundly philosophical question, to which not many would venture a reply, unless they were of a religious cast of mind, in which case they would say that God, in His wisdom, made us like that.

Why anyone should worship a Being that played such tricks on His creations is another matter.

Those who stubbornly maintain that they have no philosophy are mistaken. Nature abhors a vacuum. People who lack a coherently worked-out philosophical standpoint will inevitably reflect the ideas and prejudices of the society and the milieu in which they live. That means, in the given context, that their heads will be full of the ideas they imbibe from the newspapers, television, pulpit and schoolroom, which faithfully reflect the interests and morality of existing society.

Most people usually succeed in muddling through life, until some great upheaval compels them to re-consider the kind of ideas and values they grew up with. The crisis of society forces them to question many things they took for granted. At such times, ideas which seemed remote suddenly become strikingly relevant. Anyone who wishes to understand life, not as a meaningless series of accidents or an unthinking routine, must occupy themselves with philosophy, that is, with thought at a higher level than the immediate problems of everyday existence. Only by this means do we raise ourselves to a height where we begin to fulfill our potential as conscious human beings, willing and able to take control of our own destinies.

It is generally understood that anything worth while in life requires some effort. The study of philosophy, by its very nature, involves certain difficulties, because it deals with matters far removed from the world of ordinary experience. Even the terminology used presents difficulties because words are used in a way that does not necessarily correspond to the common usage. But the same is true for any specialized subject, from psychoanalysis to engineering.

The second obstacle is more serious. In the last century, when Marx and Engels first published their writings on dialectical materialism, they could assume that many of their readers had at least a working knowledge of classical philosophy, including Hegel. Nowadays it is not possible to make such an assumption. Philosophy no longer occupies the place it had before, since the role of speculation about the nature of the universe and life has long since been occupied by the sciences. The possession of powerful radio telescopes and spacecraft renders guesses about the nature and extent of our solar system unnecessary. Even the mysteries of the human soul are being gradually laid bare by the progress of neurobiology and psychology.

The situation is far less satisfactory in the realm of the social sciences, mainly because the desire for accurate knowledge often decreases to the degree that science impinges on the powerful material interests which govern the lives of people. The great advances made by Marx and Engels in the sphere of social and historical analysis and economics fall outside the scope of the present work.

Suffice it to point out that, despite the sustained and frequently malicious attacks to which they were subjected from the beginning, the theories of Marxism in the social sphere have been the decisive factor in the development of modern social sciences. As for their vitality, this is testified to by the fact that the attacks not only continue, but tend to increase in intensity as time goes by.

In past ages, the development of science, which has always been closely linked to that of the productive forces, had not reached a sufficiently high level to permit men and women to understand the world in which they lived. In the absence of scientific knowledge, or the material means of obtaining it, they were compelled to rely upon the one instrument they possessed that could help them to make sense of the world, and thus gain power over it — the human mind. The struggle to understand the world was closely identified with humankind's struggle to tear itself away from a merely animal level of existence, to gain mastery over the blind forces of nature, and to become free in the real, not legalistic, sense of the word. This struggle is a red thread running through the whole of human history.

Role of Religion

> *"Man is quite insane. He wouldn't know how to create a maggot, and he creates Gods by the dozen."* (Montaigne)

> *"All mythology overcomes and dominates and shapes the force of nature in the imagination and by the imagination; it therefore vanishes with the advent of real mastery over them."* (Marx)

Animals have no religion, and in the past it was said that this constituted the main difference between humans and "brutes." But that is just another way of saying that only humans possess consciousness in the full sense of the word. In recent years, there has been a reaction against the idea of Man as a special and unique Creation. This is undoubtedly correct, in the sense that humans developed from animals, and, in many important respects, remain animals. Not only do we share many of the bodily functions with other animals, but the genetic difference between humans and chimpanzees is less than two percent. That is a crushing answer to the nonsense of the Creationists.

Recent research with bonobo chimpanzees has proven beyond doubt that the primates closest to humans are capable of a level of mental activity similar in some respects to that of a human child. That is striking proof of the kinship between humans and the highest primates, but here the analogy begins to break down. Despite all the efforts of experimenters, captive bonobos have not been able to speak or fashion a stone tool remotely similar to the simplest implements created by early hominids. The two percent genetic difference between humans and chimpanzees marks the qualitative leap from the animal to the human. This was accomplished, not by a Creator, but by the development of the brain through manual labor.

The skill to make even the simplest stone tools involves a very high level of mental ability and abstract thought. The ability to select the right sort of stone and reject others; the choice of the correct angle to strike a blow, and the use of precisely the right amount of force — these are highly complicated intellectual actions. They imply a degree of planning and foresight not found in even the most advanced primates. However, the use and manufacture of stone tools was not the result of conscious planning, but was something forced upon man's remote ancestors by necessity. It was not consciousness that created humanity, but the necessary conditions of human existence which led to an enlarged brain, speech and culture, including religion.

The need to understand the world was closely linked to the need to survive. Those early hominids who discovered the use of stone scrapers in butchering dead animals with thick hides obtained a considerable advantage over those who were denied access to this rich supply of fats and proteins. Those who perfected their stone implements and worked out where to find the best materials stood a better chance of survival than those who did not. With the development of technique came the expansion of the mind, and the need to explain the phenomena of nature which governed their lives. Over millions of years, through trial and error, our ancestors began to establish certain relations between things. They began to make abstractions, that is, to generalize from experience and practice.

For centuries, the central question of philosophy has been the relation of thinking to being. Most people live their lives quite happily without even considering this problem. They think and act, talk and work, with not the slightest difficulty. Moreover, it would not occur to them to regard as incompatible the two most basic human activities, which are in practice inseparably linked. Even the most elementary action, if we exclude simple biologically determined reactions, demands some thought. To a degree, this is true not only of humans but also of animals, such as a cat lying in wait for a mouse. In man, however, the kind of thought and planning has a qualitatively higher character than any of the mental activities of even the most advanced of the apes.

This fact is inseparably linked to the capacity for abstract thought, which enables humans to go far beyond the immediate situation given to us by our senses. We can envisage situations, not just in the past (animals also have memory, as a dog which cowers at the sight of a stick) but also the future. We can anticipate complex situations, plan and thereby determine the outcome, and to some extent determine our own destinies. Although we do not normally think about it, this represents a colossal conquest which sets humankind apart from the rest of nature. "What is distinctive of human reasoning," says Professor Gordon Childe, "is that it can go immensely farther from the actual present situation than any other animal's reasoning ever seems to get it."[6] From this capacity springs all the manifold creations of civilization, culture, art, music, literature, science, philosophy, religion. We also take for granted that all this does not drop from the skies, but is the product of millions of years of development.

The Greek philosopher Anaxagoras (500-428 B.C.), in a brilliant deduction, said that man's mental development depended upon the freeing of the hands. In his important article, *The Part Played by Labor in the Transition from Ape to Man*, Engels showed the exact way in which this transition was achieved. He proved that the upright stance, freeing of the hands for labor, the form of the hands, with the opposition of the thumb to the fingers, which allowed for clutching, were the physiological preconditions for tool making, which, in turn, was the main stimulus to the development of the brain. Speech itself, which is inseparable from thought, arose out of the demands of social production, the need to realize complicated functions by means of cooperation. These theories of Engels have been strikingly confirmed by the most recent discoveries of paleontology, which show that hominid apes appeared in Africa far earlier than previously thought, and that they had brains no bigger than those of a modern chimpanzee. That is to say, the development of the brain came after the production of tools, and as a result of it. Thus, it is not true that *"In the beginning was the Word,"* but as the German poet Goethe proclaimed — *"In the beginning was the Deed."*

The ability to engage in abstract thought is inseparable from language. The celebrated prehistorian Gordon Childe observes: "Reasoning, and all that we call thinking, including the chimpanzee's, must involve mental operations with what psychologists call images. A visual image, a mental picture of, say, a banana, is always liable to be a picture of a particular banana in a particular setting. A word on the contrary is, as explained, more general and abstract, having eliminated just those accidental features that give individuality to any real banana. Mental images of words (pictures of the sound or of the muscular movements entailed in uttering it) form very convenient counters for thinking with. Thinking with their aid necessarily possesses just that quality of abstractness and generality that animal thinking seems to lack. Men can think, as well as talk, about the class of objects called 'bananas'; the chimpanzee never gets further than 'that banana in that tube.' In this way the social instrument termed language has contributed to what is grandiloquently described as 'man's emancipation from bondage to the concrete.'"[7]

Early humans, after a long period of time, formed the general idea of, say, a plant or an animal. This arose out of the concrete observation of many particular plants and animals. But when we arrive at the general concept "plant," we no longer see before us this or that flower or bush, but that which is common to all of them. We grasp the essence of a plant, its innermost being. Compared with this, the peculiar features of individual plants seem secondary and unstable. What is permanent and universal is contained in the general conception. We can never actually see a plant as such, as opposed to particular flowers and bushes. It is an abstraction of the mind. Yet it is a deeper and truer expression of what is essential to the plant's nature when stripped of all secondary features.

However, the abstractions of early humans were far from having a scientific character. They were tentative explorations, like the impressions of a child — guesses and hypotheses, sometimes incorrect, but always bold and imaginative. To our remote ancestors, the sun was a great being that sometimes warmed them, and sometimes burnt them. The earth was a sleeping giant. Fire was a fierce animal that bit them when they touched it. Early humans experienced thunder and lightning. This must have frightened them, as it still frightens animals and people today. But, unlike animals, humans looked for a general explanation of the phenomenon. Given the lack of any scientific knowledge, the explanation was invariably a supernatural one — some god, hitting an anvil with his hammer. To our eyes, such explanations seem merely amusing, like the naïve explanations of children. Nevertheless, at this period they were extremely important hypotheses — an attempt to find a rational cause for the phenomenon, in which men distinguished between the immediate experience, and saw something entirely separate from it.

The most characteristic form of early religion is animism — the notion that everything, animate or inanimate, has a spirit. We see the same kind of reaction in a child when it smacks a table against which it has banged its head. In the same way, early humans, and certain tribes today, will ask the spirit of a tree to forgive them before cutting it down. Animism belongs to a period when humankind has not yet fully separated itself from the animal world and nature in general. The closeness of humans to the world of animals is attested to by the freshness and beauty of cave-art, where horses, deer and bison are depicted with a naturalness which can no longer be captured by the modern artist. It is the childhood of the human race, which has gone beyond recall. We can only imagine the psychology of these distant ancestors of ours. But by combining the discoveries of paleontology with anthropology, it is possible to reconstruct, at least in outline, the world from which we have emerged.

In his classic anthropological study of the origins of magic and religion, Sir James Frazer writes:

A savage hardly conceives the distinction commonly drawn by more advanced peoples between the natural and the supernatural. To him the world is to a great extent worked by supernatural agents, that is, by personal beings acting on impulses and motives like his own, liable like him to be moved by appeals to their pity, their hope, and their fears. In a world so conceived he sees no limit to this power of influencing the course of nature to his own advantage. Prayers, promises, or threats may secure him fine weather and an abundant crop from the gods; and if a god should happen, as he sometimes believes, to become incarnate in his own person, then he need appeal to no higher being; he, the savage, possesses in himself all the powers necessary to further his own well-being and that of his fellow-men.[8]

The notion that the soul exists separate and apart from the body comes down from the most remote period of savagery. The basis of it is quite clear. When we are asleep, the soul appears to leave the body and roam about in dreams. By extension, the similarity between death and sleep ("death's second self," Shakespeare called it) suggested the idea that the soul could continue to exist after death. Early humans thus concluded that there is something inside them that is separate from their bodies. This is the soul, which commands the body, and can do all kinds of incredible things, even when the body is asleep. They also noticed how words of wisdom issued from the mouths of old people, and concluded that, whereas the body perishes, the soul lives on. To people used to the idea of migration, death was seen as the migration of the soul, which needed food and implements for the journey.

At first these spirits had no fixed abode. They merely wandered about, usually making trouble, which obliged the living to go to extraordinary lengths to appease them. Here we have the origin of religious ceremonies. Eventually, the idea arose that the assistance of these spirits could be enlisted by means of prayer. At this stage, religion (magic), art and science were not differentiated. Lacking the means to gain real power over their environment, early humans attempted to obtain their ends by means of magical intercourse with nature, and thus subject it to their will. The attitude of early humans to their spirit-gods and fetishes was quite practical. Prayers were intended to get results. A man would make an image with his own hands, and prostrate himself before it. But if the desired result was not forthcoming, he would curse it and beat it, in order to extract by violence what he failed to do by entreaty. In this strange world of dreams and ghosts, this world of religion, the primitive mind saw every happening as the work of unseen spirits. Every bush and stream was a living creature, friendly or hostile. Every chance event, every dream, pain or sensation, was caused by a spirit. Religious explanations filled the gap left by lack of knowledge of the laws of nature. Even death was not seen as a natural occurrence, but a result of some offence caused to the gods.

For the great majority of the existence of the human race, the minds of men and women have been full of this kind of thing. And not only in what people like to regard as primitive societies. The same kind of superstitious beliefs continue to exist in slightly different guises today. Beneath the thin veneer of civilization lurk primitive irrational tendencies and ideas which have their roots in a remote past which has been half-forgotten, but is not yet overcome. Nor will they be finally rooted out of human consciousness until men and women establish firm control over their conditions of existence.

Division of Labor

Frazer points out that the division between manual and mental labor in primitive society is invariably linked to the formation of a caste of priests, shamans or magicians:

> Social progress, as we know, consists mainly in a successive differentiation of functions, or, in simpler language, a division of labor. The work which in primitive society is done by all alike and by all equally ill, or nearly so, is gradually distributed among different classes of workers and executed more and more perfectly; and so far as the products, material or immaterial, of his specialized labor are shared by all, the whole community benefits by the increasing specialization. Now magicians or medicine-men appear to constitute the oldest artificial or professional class in the evolution of society. For sorcerers are found in every savage tribe known to us; and among the lowest savages, such as the Australian aborigines, they are the only professional class that exists. [9]

The dualism which separates soul from body, mind from matter, thinking from doing, received a powerful impulse from the development of the division of labor at a given stage of social evolution. The separation between mental and manual labor is a phenomenon which coincides with the division of society into classes. It marked a great advance in human development. For the first time, a minority of society was freed from the necessity to work to obtain the essentials of existence. The possession of that most precious commodity, leisure, meant that men could devote their lives to the study of the stars. As the German materialist philosopher Ludwig Feuerbach explains, real theoretical science begins with cosmology:

> The animal is sensible only of the beam which immediately affects life; while man perceives the ray, to him physically indifferent, of the remotest star. Man alone has purely intellectual, disinterested joys and passions; the eye of man alone keeps theoretic festivals. The eye which looks into the starry heavens, which gazes at that light, alike useless and harmless, having nothing in common with the earth and its necessities - this eye sees in that light its own nature, its own origin. The eye is heavenly in its nature. Hence man elevates himself above the earth only with the

eye; hence theory begins with the contemplation of the heavens. The first philosophers were astronomers.[10]

Although at this early stage this was still mixed up with religion, and the requirements and interests of a priest caste, it also signified the birth of human civilization. This was already understood by Aristotle, who wrote:

"These theoretical arts, moreover, were evolved in places where men had plenty of free time: mathematics, for example, originated in Egypt, where a priestly caste enjoyed the necessary leisure."[11]

Knowledge is a source of power. In any society in which art, science and government is the monopoly of a few, that minority will use and abuse its power in its own interests. The annual flooding of the Nile was a matter of life and death to the people of Egypt, whose crops depended on it. The ability of the priests in Egypt to predict, on the basis of astronomical observations, when the Nile would flood its banks must have greatly increased their prestige and power over society. The art of writing, a most powerful invention, was the jealously guarded secret of the priest-caste. As Ilya Prigogine and Isabelle Stengers comment:

"Sumer discovered writing; the Sumerian priests speculated that the future might be written in some hidden way in the events taking place around us in the present. They even systematized this belief, mixing magical and rational elements."[12]

The further development of the division of labor gave rise to an unbridgeable gulf between the intellectual elite and the majority of humankind, condemned to labor with their hands. The intellectual, whether Babylonian priest or modern theoretical physicist, knows only one kind of labor, mental labor. Over the course of millennia, the superiority of the latter over "crude" manual labor becomes deeply ingrained and acquires the force of a prejudice. Language, words and thoughts become endowed with mystical powers. Culture becomes the monopoly of a privileged elite, which jealously guards its secrets, and uses and abuses its position in its own interests.

In ancient times, the intellectual aristocracy made no attempt to conceal its contempt for physical labor. The following extract from an Egyptian text known as *The Satire on the Trades*, written about 2000 B.C. is supposed to consist of a father's exhortation to his son, whom he is sending to the Writing School to train as a scribe:

I have seen how the belabored man is belabored — thou shouldst set thy heart in pursuit of writing. And I have observed how one may be rescued from his duties [sic!] — behold, there is nothing which surpasses writing...

I have seen the metalworker at his work at the mouth of his furnace. His fingers were somewhat like crocodiles; he stank more than fish-roe...

The small building contractor carries mud...He is dirtier than vines or pigs from treading under his mud. His clothes are stiff with clay...

The arrow-maker, he is very miserable as he goes out into the desert [to get flint points]. Greater is that which he gives to his donkey than its work thereafter [is worth]...

The laundry man launders on the [river] bank, a neighbor of the crocodile...

Behold, there is no profession free of a boss — except for the scribe: he is the boss...

Behold, there is no scribe who lacks food from the property of the House of the King — life, prosperity, health!...His father and his mother praise god, he being set upon the way of the living. Behold these things — I [have set them] before thee and thy children's children. [13]

The same attitude was prevalent among the Greeks:

What are called the mechanical arts [says Xenophon], carry a social stigma and are rightly dishonored in our cities, for these arts damage the bodies of those who work in them or who act as overseers, by compelling them to a sedentary life and to an indoor life, and, in some cases, to spend the whole day by the fire. This physical degeneration results also in deterioration of the soul. Furthermore, the workers at these trades simply have not got the time to perform the offices of friendship or citizenship. Consequently they are looked upon as bad friends and bad patriots, and in some cities, especially the warlike ones, it is not legal for a citizen to ply a mechanical trade. [14]

The radical divorce between mental and manual labor deepens the illusion that ideas, thoughts and words have an independent existence. This misconception lies at the heart of all religion and philosophical idealism.

It was not god who created man after his own image, but, on the contrary, men and women who created gods in their own image and likeness. Ludwig Feuerbach said that if birds had a religion, their God would have wings. "Religion is a dream, in which our own conceptions and emotions appear to us as separate existences, beings out of ourselves. The religious mind does not distinguish between subjective and objective — it has no doubts; it has the faculty, not of discerning other things than itself, but of seeing its own conceptions out of itself as distinct beings." [15] This was already understood by men like Xenophanes of Colophon (565-c.470 B.C.), who wrote, "Homer and Hesiod have ascribed to the gods every deed that is shameful and dishonorable among men: stealing and adultery and deceiving each other...The Ethiopians make their gods black and snub-nosed, and the Thracians theirs grey-eyed and red-haired...If animals could paint and make things, like men, horses and oxen too would fashion the gods in their own image." [16]

The Creation myths which exist in almost all religions invariably take their images from real life, for example, the image of the potter who gives form to formless clay. In the opinion of Gordon Childe, the story of the Creation in the first book of *Genesis* reflects the fact that, in Mesopotamia the land was indeed separated from the waters *"in the Beginning,"* but not by divine intervention:

> The land on which the great cities of Babylonia were to rise had literally to be created; the prehistoric forerunner of the biblical Erech was built on a sort of platform of reeds, laid criss-cross upon the alluvial mud. The Hebrew book of Genesis has familiarized us with much older traditions of the pristine condition of Sumer — a "chaos" in which the boundaries between water and dry land were still fluid. An essential incident in "The Creation" is the separation of these elements. Yet it was no god, but the proto-Sumerians themselves who created the land; they dug channels to water the fields and drain the marsh; they built dykes and mounded platforms to protect men and cattle from the waters and raise them above the flood; they made the first clearings in the reed brakes and explored the channels between them. The tenacity with which the memory of this struggle persisted in tradition is some measure of the exertion imposed upon the ancient Sumerians. Their reward was an assured supply of nourishing dates, a bounteous harvest from the fields they had drained, and permanent pastures for flocks and herds. [17]

Man's earliest attempts to explain the world and his place in it were mixed up with mythology. The Babylonians believed that the god Marduk created Order out of Chaos, separating the land from the water, heaven from earth. The biblical Creation myth was taken from the Babylonians by the Jews, and later passed into the culture of Christianity. The true history of scientific thought commences when men and women learn to dispense with mythology, and attempt to obtain a rational understanding of nature, without the intervention of the gods. From that moment, the real struggle for the emancipation of humanity from material and spiritual bondage begins.

The advent of philosophy represents a genuine revolution in human thought. Like so much of modern civilization, we owe it to the ancient Greeks. Although important advances were also made by the Indians and Chinese, and later the Arabs, it was the Greeks who developed philosophy and science to its highest point prior to the Renaissance. The history of Greek thought in the four hundred year period, from the middle of the 7th century B.C., constitutes one of the most imposing pages in the annals of human history.

Materialism and Idealism

The whole history of philosophy from the Greeks down to the present day consist of a struggle between two diametrically opposed schools of thought — materialism and idealism. Here we come across a perfect example of how the terms used in philosophy differ fundamentally from everyday language.

When we refer to someone as an *"idealist"* we normally have in mind a person of high ideals and spotless morality. A materialist, on the contrary, is viewed as an unprincipled so-and-so, a money-grubbing, self-centered individual with gross appetites for food and other things — in short, a thoroughly undesirable character.

This has nothing whatever to do with philosophical materialism and idealism. In a philosophical sense, idealism sets out from the view that the world is only a reflection of ideas, mind, spirit, or more correctly the Idea, which existed before the physical world. The crude material things we know through our senses are, according to this school, only imperfect copies of this perfect Idea. The most consistent proponent of this philosophy in Antiquity was Plato. However, he did not invent idealism, which existed earlier.

The Pythagoreans believed that the essence of all things was Number (a view apparently shared by some modern mathematicians). The Pythagoreans displayed a contempt towards the material world in general and the human body in particular which they saw as a prison where the soul was trapped. This is strikingly reminiscent of the outlook of mediaeval monks. Indeed, it is probable that the Church took many of its ideas from the Pythagoreans, Platonists and Neo-Platonists. This is not surprising. All religions necessarily set out from an idealist view of the world. The difference is that religion appeals to the emotions, and claims to provide a mystical, intuitive understanding of the world (*"Revelation"*), while most idealist philosophers try to present logical arguments for their theories.

At bottom, however, the roots of all forms of idealism are religious and mystical. The disdain for the "crude material world" and the elevation of the "Ideal" flow directly from the phenomena we have just considered in relation to religion. It is no accident that Platonist idealism developed in Athens when the system of slavery was at its height. Manual labor at that time was seen, in a very literal sense, as a mark of slavery. The only labor worthy of respect was intellectual labor. Essentially, philosophical idealism is a product of the extreme division between mental and manual labor which has existed from the dawn of written history down to the present day.

The history of Western philosophy, however, begins not with idealism but with materialism. This asserts precisely the opposite: that the material world, known to us and explored by science, is real; that the only real world is the material one; that thoughts, ideas and sensations are the product of matter organized in a certain way (a nervous system and a brain); that thought cannot derive its categories from itself, but only from the objective world which makes itself known to us through our senses.

The earliest Greek philosophers were known as "hylozoists" (from the Greek, meaning "those who believe that matter is alive"). Here we have a long line of heroes who pioneered the development of thought. The Greeks discovered that the world was round, long before Columbus. They explained that humans had evolved from fishes long before Darwin. They made extraordinary discoveries in mathematics, especially geometry, which were not greatly improved upon for one and a half millennia. They invented mechanics and even built a steam engine. What was startlingly new about this way of looking at the world was that it was not religious. In complete contrast to the Egyptians and Babylonians, from whom they had learnt a lot, the Greek thinkers did not resort to gods and goddesses to explain natural phenomena. For the first time, men and women sought to explain the workings of nature purely in terms of nature. This was one of the greatest turning points in the entire history of human thought. True science starts here.

Aristotle, the greatest of the Ancient philosophers, can be considered a materialist, although he was not so consistent as the early hylozoists. He made a series of important scientific discoveries which laid the basis for the great achievements of the Alexandrine period of Greek science.

The Middle Ages which followed the collapse of Antiquity were a desert in which scientific thought languished for centuries. Not accidentally, this was a period dominated by the Church. Idealism was the only philosophy permitted, either as a caricature of Plato or an even worse distortion of Aristotle.

Science re-emerged triumphantly in the period of the Renaissance. It was forced to wage a fierce battle against the influence of religion (not only Catholic, but also Protestant, by the way). Many martyrs paid the price of scientific freedom with their lives. Giordano Bruno was burnt at the stake. Galileo was twice put on trial by the Inquisition, and forced to renounce his views under threat of torture.

The predominant philosophical trend of the Renaissance was materialism. In England, this took the form of empiricism, the school of thought that states that all knowledge is derived from the senses. The pioneers of this school were Francis Bacon (1561-1626), Thomas Hobbes (1588-1679) and John Locke (1632-1704). The materialist school passed from England to France where it acquired a revolutionary content. In the hands of Diderot, Rousseau, Holbach and Helvetius, philosophy became an instrument for criticizing all existing society. These great thinkers prepared the way for the revolutionary overthrow of the feudal monarchy in 1789-93.

The new philosophical views stimulated the development of science, encouraging experiment and observation. The 18th century saw a great advance in science, especially mechanics. But this fact had a negative as well as a positive side. The old materialism of the 18th century was narrow and rigid, reflecting the limited development of science itself. Newton expressed the limitations of empiricism with his celebrated phrase "I make no hypotheses." This one-sided mechanical outlook ultimately proved fatal to the old materialism. Paradoxically, the great advances in philosophy after 1700 were made by idealist philosophers.

Under the impact of the French revolution, the German idealist Immanuel Kant (1724-1804) subjected all previous philosophy to a thorough criticism. Kant made important discoveries not only in philosophy and logic but in science. His nebular hypothesis of the origins of the solar system (later given a mathematical basis by Laplace) is now generally accepted as correct. In the field of philosophy, Kant's masterpiece *The Critique of Pure Reason* was the first work to analyze the forms of logic which had remained virtually unchanged since they were first developed by Aristotle. Kant showed the contradictions implicit in many of the most fundamental propositions of philosophy. However, he failed to resolve these contradictions ("*Antinomies*"), and finally drew the conclusion that real knowledge of the world was impossible. While we can know appearances, we can never know how things are "in themselves."

This idea was not new. It is a theme which has recurred many times in philosophy, and is generally identified with what we call subjective idealism. This was put forward before Kant by the Irish bishop and philosopher George Berkeley and the last of the classical English empiricists, David Hume. The basic argument can be summed up as follows: "I interpret the world through my senses. Therefore, all that I know to exist are my sense-impressions. Can I, for example, assert that this apple exists? No. All I can say is that I see it, I feel it, I smell it, I taste it. Therefore, I cannot really say that the material world exists at all." The logic of subjective idealism is that, if I close my eyes, the world ceases to exist. Ultimately, it leads to solipsism (from the Latin "*solo ipsus*" — "*I alone*"), the idea that only I exist.

These ideas may seem nonsensical to us, but they have proved strangely persistent. In one way or another, the prejudices of subjective idealism have penetrated not only philosophy but also science for a great part of the 20th century. We shall deal more specifically with this trend later on.

The greatest breakthrough came in the first decades of the 19th century with George Wilhelm Friedrich Hegel (1770-1831). Hegel was a German idealist, a man of towering intellect, who effectively summed up in his writings the whole history of philosophy.

Hegel showed that the only way to overcome the "Antinomies" of Kant was to accept that contradictions actually existed, not only in thought, but in the real world. As an objective idealist, Hegel had no time for the subjective idealist argument that the human mind cannot know the real world. The forms of thought must reflect the objective world as closely as possible. The process of knowledge consist of penetrating ever more deeply into this reality, proceeding from the abstract to the concrete, from the known to the unknown, from the particular to the universal.

The dialectical method of thinking had played a great role in Antiquity, particularly in the naïve but brilliant aphorisms of Heraclitus (c. 500 B.C.), but also in Aristotle and others. It was abandoned in the Middle Ages, when the Church turned Aristotle's formal logic into a lifeless and rigid dogma, and did not re-appear until Kant returned it to a place of honor. However, in Kant the dialectic did not receive an adequate development. It fell to Hegel to bring the science of dialectical thinking to its highest point of development.

Hegel's greatness is shown by the fact that he alone was prepared to challenge the dominant philosophy of mechanism. The dialectical philosophy of Hegel deals with processes, not isolated events. It deals with things in their life, not their death, in their inter-relations, not isolated, one after the other. This is a startlingly modern and scientific way of looking at the world. Indeed, in many aspects Hegel was far in advance of his time. Yet, despite its many brilliant insights, Hegel's philosophy was ultimately unsatisfactory. Its principal defect was precisely Hegel's idealist standpoint, which prevented him from applying the dialectical method to the real world in a consistently scientific way. Instead of the material world we have the world of the Absolute Idea, where real things, processes and people are replaced by insubstantial shadows. In the words of Frederick Engels, the Hegelian dialectic was the most colossal miscarriage in the whole history of philosophy. Correct ideas are here seen standing on their head. In order to put dialectics on a sound foundation, it was necessary to turn Hegel upside down, to transform idealist dialectics into dialectical materialism. This was the great achievement of Karl Marx and Frederick Engels. Our study begins with a brief account of the basic laws of materialist dialectics worked out by them.

3.   DIALECTICAL MATERIALISM

What is Dialectics?

> "Panta cwrei, oudei menei."
> "Everything flows and nothing stays." (Heraclitus)

Dialectics is a method of thinking and interpreting the world of both nature and society. It is a way of looking at the universe, which sets out from the axiom that everything is in a constant state of change and flux. But not only that. Dialectics explains that change and motion involve contradiction and can only take place through contradictions. So instead of a smooth, uninterrupted line of progress, we have a line which is interrupted by sudden and explosive periods in which slow, accumulated changes (quantitative change) undergoes a rapid acceleration, in which quantity is transformed into quality. Dialectics is the logic of contradiction.

The laws of dialectics were already worked out in detail by Hegel, in whose writings, however, they appear in a mystified, idealist form. It was Marx and Engels who first gave dialectics a scientific, that is to say, materialist basis. "Hegel wrote before Darwin and before Marx," wrote Trotsky. "Thanks to the powerful impulse given to thought by the French Revolution, Hegel anticipated the general movement of science. But because it was only an anticipation, although by a genius, it received from Hegel an idealistic character. Hegel operated with ideological shadows as the ultimate reality. Marx demonstrated that the movement of these ideological shadows reflected nothing but the movement of material bodies." [18]

In the writings of Hegel there are many striking examples of the law of dialectics drawn from history and nature. But Hegel's idealism necessarily gave his dialectics a highly abstract, and arbitrary character. In order to make dialectics serve the "Absolute Idea," Hegel was forced to impose a schema upon nature and society, in flat contradiction to the dialectical method itself, which demands that we derive the laws of a given phenomenon from a scrupulously objective study of the subject-matter as Marx did in his *Capital*. Thus, far from being a mere regurgitation of Hegel's idealist dialectic arbitrarily foisted on history and society as his critics often assert, Marx's method was precisely the opposite. As he himself explains:

> My dialectic method [wrote Marx], is not only different from the Hegelian, but is its direct opposite. To Hegel, the life-process of the human brain, i.e. the process of thinking, which, under the name of 'the Idea,' he even transforms into an independent subject, is the demiurgos of the real world, and the real world is only the external, phenomenal form of "the Idea." With me, on the contrary, the ideal is nothing else than the material world reflected by the human mind, and translated into forms of thought. [19]

When we first contemplate the world around us, we see an immense and amazingly complex series of phenomena, an intricate web of seemingly endless change, cause and effect, action and reaction. The motive force of scientific investigation is the desire to obtain a rational insight into this bewildering labyrinth, to understand it in order to conquer it. We look for laws which can separate the general from the particular, the accidental from the necessary, and enable us to understand the forces that give rise to the phenomena which confront us.

In the words of the English physicist and philosopher David Bohm:

> In nature nothing remains constant. Everything is in a perpetual state of transformation, motion, and change. However, we discover that nothing simply surges up out of nothing without having antecedents that existed before. Likewise, nothing ever disappears without a trace, in the sense that it gives rise to absolutely nothing existing at later times. This general characteristic of the world can be expressed in terms of a principle which summarizes an enormous domain of different kinds of experience and which has never yet been contradicted in any observation or experiment, scientific or otherwise; namely, everything comes from other things and gives rise to other things.[20]

The fundamental proposition of dialectics is that everything is in a constant process of change, motion and development. Even when it appears to us that nothing is happening, in reality, matter is always changing. Molecules, atoms and subatomic particles are constantly changing place, always on the move. Dialectics is thus an essentially dynamic interpretation of the phenomena and processes which occur at all levels of both organic and inorganic matter.

> To our eyes, our crude eyes, nothing is changing," notes the American physicist Richard P. Feynman, "but if we could see it a billion times magnified, we would see that from its own point of view it is always changing: molecules are leaving the surface, molecules are coming back.[21]

So fundamental is this idea to dialectics that Marx and Engels considered motion to be the most basic characteristic of matter. As in so many cases, this dialectical notion was already anticipated by Aristotle, who wrote: "Therefore...the primary and proper meaning of 'nature' is the essence of things which have in themselves...the principle of motion." [22] This is not the mechanical conception of motion as something imparted to an inert mass by an external "force" but an entirely different notion of matter as self-moving. For them, matter and motion (energy) were one and the same thing, two ways of expressing the same idea. This idea was brilliantly confirmed by Einstein's theory of the equivalence of mass and energy. This is how Engels expresses it:

> Motion in the most general sense, conceived as the mode of existence, the inherent attribute, of matter, comprehends all changes and processes occurring in the universe, from mere change of place right up to thinking. The investigation of the nature of motion had as a matter of course to start from the lowest, simplest forms of this motion and to learn to grasp these before it could achieve anything in the way of explanation of the higher and more complicated forms. [23]

"Everything Flows"

Everything is in a constant state of motion, from neutrinos to superclusters. The earth itself is constantly moving, rotating around the sun once a year, and rotating on its own axis once a day. The sun, in turn, revolves on its axis once in 26 days and, together with all the other stars in our galaxy, travels once around the galaxy in 230 million years. It is probable that still larger structures (clusters of galaxies) also have some kind of overall rotational motion. This seems to be a characteristic of matter right down to the atomic level, where the atoms which make up molecules rotate about each other at varying rates. Inside the atom, electrons rotate around the nucleus at enormous speeds.

The electron possesses a quality known as intrinsic spin. It is as if it rotates around its own axis at a fixed rate and cannot be stopped or changed except by destroying the electron as such. If the spin of the electron is increased, it so drastically alters its properties that it results in a qualitative change, producing a completely different particle. The quantity known as angular momentum — the combined measure of the mass, size and speed of the rotating system — is used to measure the spin of elementary particles. The principle of spin quantization is fundamental at the subatomic level but also exists in the macroscopic world. However, its effect is so infinitesimal that it can be taken for granted. The world of subatomic particles is in a state of constant movement and ferment, in which nothing is ever the same as itself. Particles are constantly changing into their opposites, so that it is impossible even to assert their identity at any given moment of time. Neutrons change into protons, and protons into neutrons in a ceaseless exchange of identity.

Engels defines dialectics as "the science of the general laws of motion and development of nature, human society and thought." In *Anti-Dühring* and *The Dialectics of Nature*, Engels gives an account of the laws of dialectics, beginning with the three most fundamental ones:

1) The law of the transformation of quantity into quality and vice versa;
2) The law of the interpenetration of opposites, and
3) The law of the negation of the negation.

At first sight, such a claim may seem excessively ambitious. Is it really possible to work out laws which have such a general application? Can there be an underlying pattern which repeats itself in the workings, not only of society and thought, but of nature itself? Despite all such objections, it is becoming increasingly clear that such patterns do indeed exist and constantly re-appear at all kinds of levels, in all kinds of ways. And there is an increasing number of examples, drawn from fields as diverse as subatomic particles to population studies, which lend increasing weight to the theory of dialectical materialism.

The essential point of dialectical thought is not that it is based on the idea of change and motion but that it views motion and change as phenomena based upon contradiction. Whereas traditional formal logic seeks to banish contradiction, dialectical thought embraces it. Contradiction is an essential feature of all being. It lies at the heart of matter itself. It is the source of all motion, change, life and development. The dialectical law which expresses this idea is the law of the unity and interpenetration of opposites. The third law of dialectics, the negation of the negation, expresses the notion of development. Instead of a closed circle, where processes continually repeat themselves, this law points out that movement through successive contradictions actually leads to development, from simple to complex, from lower to higher. Processes do not repeat themselves exactly in the same way, despite appearances to the contrary. These, in a very schematic outline, are the three most fundamental dialectical laws. Arising from them there are a whole series of additional propositions, involving the relation between whole and part, form and content, finite and infinite, attraction and repulsion and so on. These we shall attempt to deal with. Let us begin with quantity and quality.

Quantity and Quality

The law of the transformation of quantity into quality has an extremely wide range of applications, from the smallest particles of matter at the subatomic level to the largest phenomena known to man. It can be seen in all kinds of manifestations, and at many levels. Yet this very important law has yet to receive the recognition which it deserves. This dialectical law forces itself to our attention at every turn. The transformation of quantity into quality was already known to the Megaran Greeks, who used it to demonstrate certain paradoxes, sometimes in the form of jokes. For example, the "bald head" and the "heap of grain" — does one hair less mean a bald head, or one grain of corn a heap? The answer is no. Nor one more? The answer is still no. The question is then repeated until there is a heap of corn and a bald head. We are faced with the contradiction that the individual small changes, which are powerless to effect a qualitative change, at a certain point do exactly that: quantity changes into quality.

The idea that, under certain conditions, even small things can cause big changes finds its expression in all kinds of sayings and proverbs. For instance: "The straw that broke the camel's back," "many hands make light work," "constant dripping wears away the stone," and so on. In many ways, the law of the transformation of quantity into quality has penetrated the popular consciousness, as Trotsky wittily pointed out:

"Every individual is a dialectician to some extent or other, in most cases, unconsciously. A housewife knows that a certain amount of salt flavors soup agreeably, but that added salt makes the soup unpalatable. Consequently, an illiterate peasant woman guides herself in cooking soup by the Hegelian law of the transformation of quantity into quality. Similar examples from daily life could be cited without end. Even animals arrive at their practical conclusions not only on the basis of the Aristotelian syllogism but also on the basis of the Hegelian dialectic. Thus a fox is aware that quadrupeds and birds are nutritious and tasty. On sighting a hare, a rabbit, or a hen, a fox concludes: this particular creature belongs to the tasty and nutritive type, and — chases after the prey. We have here a complete syllogism, although the fox, we may suppose, never read Aristotle. When the same fox, however, encounters the first animal which exceeds it in size, for example, a wolf, it quickly concludes that quantity passes into quality, and turns to flee. Clearly, the legs of a fox are equipped with Hegelian tendencies, even if not fully conscious ones.

> All this demonstrates, in passing, that our methods of thought, both formal logic and the dialectic, are not arbitrary constructions of our reason but rather expressions of the actual inter-relationships in nature itself. In this sense, the universe throughout is permeated with "unconscious" dialectics. But nature did not stop there. No little development occurred before nature's inner relationships were converted into the language of the consciousness of foxes and men, and man was then enabled to generalize these forms of consciousness and transform them into logical (dialectical) categories, thus creating the possibility for probing more deeply into the world about us. [24]

Despite the apparently trivial character of these examples, they do reveal a profound truth about the way the world works. Take the example of the heap of corn. Some of the most recent investigations related to chaos theory have centered on the critical point where a series of small variations produces a massive change of state. (In the modern terminology, this is called "the edge of chaos.") The work of the Danish-born physicist Per Bak and others on "self-organized criticality" used precisely the example of a sand-heap to illustrate profound processes which occur at many levels of nature and which correspond precisely to the law of the transformation of quantity into quality.

One of the examples of this is that of a pile of sand — a precise analogy with the heap of grain of the Megarans. We drop grains of sand one by one on a flat surface. The experiment has been conducted many times, both with real sand heaped on tables, and in computer simulations. For a time they will just pile up on top of each other until they make a little pyramid. Once this point is reached, any additional grains will either find a resting place on the pile, or will unbalance one side of it just enough to cause some of the other grains to fall in an avalanche. Depending on how the other grains are poised, the avalanche could be very small, or devastating, dragging a large number of grains with it. When the pile reaches this critical point, even a single grain would be capable of dramatically affecting all around it. This seemingly trivial example provides an excellent "edge-of-chaos model," with a wide range of applications, from earthquakes to evolution; from stock exchange crises to wars.

The pile of sand grows bigger, with excess sand slipping from the sides. When all the excess sand has fallen off, the resulting sand-pile is said to be "self-organized." In other words, no-one has consciously shaped it in this way. It "organizes itself," according to its own inherent laws, until it reaches a state of criticality, in which the sand grains on its surface are barely stable. In this critical condition, even the addition of a single grain of sand can cause unpredictable results. It may just cause a further tiny shift, or it may trigger a chain-reaction resulting in a catastrophic landslide and the destruction of the pile.

According to Per Bak, the phenomenon can be given a mathematical expression, according to which the average frequency of a given size of avalanche is inversely proportional to some power of its size. He also points out that this "power-law" behavior is extremely common in nature, as in the critical mass of plutonium, at which the chain-reaction is on the point of running away into a nuclear explosion. At the sub-critical level, the chain-reaction within the plutonium mass will die out, whereas a supercritical mass will explode. A similar phenomenon can be seen in earthquakes, where the rocks on two sides of a fault in the earth's crust reach a point where they are ready to slip past each other. The fault experiences a series of little slips and bigger slips, which maintain the tension at the critical point for some time until it finally collapses into an earthquake. Although the proponents of chaos theory seem unaware of it, these examples are all cases of the law of the transformation of quantity into quality. Hegel invented the nodal line of measure relations, in which small quantitative changes at a certain point give rise to a qualitative leap. The example is often given of water, which boils at 100°C at normal atmospheric pressure. As the temperature nears boiling point, the increase in heat does not immediately cause the water molecules to fly apart. Until it reaches boiling point, the water keeps its volume. It remains water, because of the attraction of the molecules for each other. However, the steady change in temperature has the effect of increasing the motion of the molecules. The volume between the atoms is gradually increased, to the point where the force of attraction is insufficient to hold the molecules together. At precisely 100°C, any increase in heat energy will cause the molecules to fly apart, producing steam.

The same process can be seen in reverse. When water is cooled from 100°C to 0°C, it does not gradually congeal, passing from a paste, through a jelly, to a solid state. The motion of the atoms is gradually slowed as heat energy is removed until, at 0°C, a critical point is reached, at which the molecules will lock into a certain pattern, which is ice. The qualitative difference between a solid and a liquid can be readily understood by anyone. Water can be used for certain purposes, like washing and quenching one's thirst, which ice cannot. Technically speaking, the difference is that, in a solid, the atoms are arranged in a crystalline array. They do not have a random position at long distances, so that the position of the atoms on one side of the crystal is determined by the atoms on the other side. That is why we can move our hand freely through water, whereas ice is rigid and offers resistance. Here we are describing a qualitative change, a change of state, which arises from an accumulation of quantitative changes. A water molecule is a relatively simple affair, one oxygen atom attached to two hydrogen atoms governed by well understood equations of atomic physics. However, when a very large number of these molecules are combined, they acquire a property which none of them possesses in isolation — liquidity. Such a property is not implied in the equations. In the language of complexity, liquidity is an "emergent" phenomenon.

"Cool those liquid water molecules down a bit, for example, and at 32°F they will suddenly quit tumbling over one another at random. Instead they will undergo a 'phase transition,' locking themselves into the orderly crystalline array known as ice. Or if you were to go the other direction and heat the liquid, those same tumbling water molecules will suddenly fly apart and undergo a phase transition into water vapor. Neither phase transition would have any meaning for one molecule alone." [25]

The phrase "phase transition" is neither more nor less than a qualitative leap. Similar processes can be seen in phenomena as varied as the weather, DNA molecules, and the mind itself. This quality of liquidity is well known on the basis of our daily experience. In physics, too, the behavior of liquids is well understood and perfectly predictable — up to a point. The laws of motion of fluids (gases and liquids) clearly distinguish between smooth laminar flow, which is well defined and predictable, and turbulent flow, which can be expressed, at best, approximately. The movement of water around a pier in a river can be accurately predicted from the normal equations for fluids, provided it is moving slowly. Even if we increase the speed of the flow, causing eddies and vortices, we can still predict their behavior. But if the speed is increased beyond a certain point, it becomes impossible to predict where the eddies will form, or, indeed, to say anything about the behavior of the water at all. It has become chaotic.

Mendeleyev's Periodic Table

The existence of qualitative changes in matter was known long before human beings began to think about science, but it was not really understood until the advent of atomic theory. Earlier, physics took the changes of state from solid to liquid to gas as something that occurred, without knowing exactly why. Only now are these phenomena being properly understood.

The science of chemistry made great strides forward in the 19th century. A large number of elements was discovered. But, rather like the confused situation which exists in particle physics today, chaos reigned. Order was established by the great Russian scientist Dimitri Ivanovich Mendeleyev who, in 1869, in collaboration with the German chemist Julius Meyer, worked out the periodic table of the elements, so-called because it showed the periodic recurrence of similar chemical properties.

The existence of atomic weight was discovered in 1862 by Cannizzaro. But Mendeleyev's genius consisted in the fact that he did not approach the elements from a purely quantitative standpoint, that is, he did not see the relation between the different atoms just in terms of weight. Had he done so, he would never have made the breakthrough he did. From the purely quantitative standpoint, for instance, the element tellurium (atomic weight = 127.61) ought to have come after iodine (atomic weight = 126.91) in the periodic table, yet Mendeleyev placed it before iodine, under selenium, to which it is more similar, and placed iodine under the related element, bromine. Mendeleyev's method was vindicated in the 20th century, when the investigation of X-rays proved that his arrangement was the correct one. The new atomic number for tellurium was put at 52, while that of iodine is 53.

The whole of Mendeleyev's periodic table is based on the law of quantity and quality, deducing qualitative differences in the elements from quantitative differences in atomic weights. This was recognized by Engels at the time:

> Finally, the Hegelian law is valid not only for compound substances but also for the chemical elements themselves. We now know that "the chemical properties of the elements are a periodic function of their atomic weights,"...and that, therefore, their quality is determined by the quantity of their atomic weight. And the test of this has been brilliantly carried out. Mendeleyev proved that various gaps occur in the series of related elements arranged according to atomic weights indicating that here new elements remain to be discovered. He described in advance the general chemical properties of one of these unknown elements, which he termed eka-aluminum, because it follows after aluminum in the series beginning with the latter, and he predicted its approximate specific and atomic weight as well as its atomic volume. A few years later, Lecoq de Boisbaudran actually discovered this element, and Mendeleyev's predictions fitted with only very slight discrepancies. Eka-aluminium was realized in gallium... By means of the — unconscious — application of Hegel's law of the transformation of quantity into quality, Mendeleyev achieved a scientific feat which it is not too bold to put on a par with that of Leverrier in calculating the orbit of the until then unknown planet Neptune.[26]

Chemistry involves changes of both a quantitative and qualitative character, both changes of degree and of state. This can clearly be seen in the change of state from gas to liquid or solid, which is usually related to variations of temperature and pressure. In *Anti Dühring*, Engels gives a series of examples of how, in chemistry, the simple quantitative addition of elements creates qualitatively different bodies. Since Engels' time the naming system used in chemistry has been changed. However, the change of quantity into quality is accurately expressed in the following example:

Table 1:

	boiling point	melting point
CH2O2 — formic acid	100°	1°
C2H4O2 acetic acid	118°	17°
C3H6O2 — propionic acid	140°	—
C4H8O2 — butyric acid	162°	—
C5H10O2 — valerianic acid	175°	—

and so on to C30H60O2, melissic acid, which melts only at 80° and has no boiling point at all, because it does not evaporate without disintegrating. [27]

The study of gases and vapors constitutes a special branch of chemistry. The great British pioneer of chemistry, Faraday, thought that it was impossible to liquefy six gases, which he called permanent gases — hydrogen, oxygen, nitrogen, carbon monoxide, nitric oxide and methane. But in 1877, the Swiss chemist R. Pictet managed to liquefy oxygen at a temperature of -140°C under a pressure of 500 atmospheres. Later, nitrogen, oxygen and carbon monoxide were all liquefied at still lower temperatures. In 1900, hydrogen was liquefied at -240° and, at a lower temperature, it even solidified. Finally, the most difficult challenge of all, the liquefaction of helium, was achieved at -255°. These discoveries had important practical applications. Liquid hydrogen and oxygen are now used in large amounts in rockets. The transformation of quantity into quality is shown by the fact that changes of temperature bring about important changes of properties. This is the key to the phenomenon of superconductivity. Through super-cooling, certain substances, beginning with mercury, were shown to offer no resistance to electric currents.

The study of extremely low temperature was developed in the mid-19th century by the Englishman William (later Lord) Kelvin, who established the concept of absolute zero (the lowest possible temperature) which he calculated to be -273°C. At this temperature, he thought, the energy of molecules would sink to zero. This temperature is sometimes referred to as zero Kelvin, and used as the basis for a scale to measure very low temperatures. However, even at absolute zero, motion is not done away with altogether. There is still some energy, which cannot be removed. For practical purposes, energy is said to be zero, but that is not actually the case. Matter and motion, as Engels pointed out, are absolutely inseparable — even at "absolute zero."

Nowadays, incredibly low temperatures are routinely achieved, and play an important role in the production of superconductors. Mercury becomes superconductive at exactly 4.12° Kelvin (K); lead at 7.22°K; tin at 3.73°K; aluminum at 1.20°K; uranium at 0.8°K, titanium at 0.53°K. Some 1,400 elements and alloys display this quality. Liquid hydrogen boils at 20.4°K. Helium is the only known substance which cannot be frozen, even at absolute zero. It is the only substance which possesses the phenomenon known as super-fluidity. Here, too, however, changes of temperature produce qualitative leaps. At 2.2°K, the behavior of helium undergoes so fundamental a change, that it is known as helium-2, to distinguish it from liquid helium above this temperature (helium-1). Using new techniques, temperatures as low as 0.000001°K have been reached, though it is thought that absolute zero is unattainable.

So far, we have concentrated on chemical changes in the laboratory and in industry. But it should not be forgotten that these changes take place on a much vaster scale in nature. The chemical composition of coal and diamonds, barring impurities, is the same — carbon. The difference is the result of colossal pressure which, at a certain point, transforms the contents of the coal-sack into a duchess' necklace. To convert common graphite into diamonds would require the pressure of at least 10,000 atmospheres over a very long period of time. This process occurs naturally beneath the earth's surface. In 1955, the big monopoly GEC succeeded in changing graphite into diamonds with a temperature of 2,500°C, and a pressure of 100,000 atmospheres. The same result was obtained in 1962, with a temperature of 5,000°C, and a pressure of 200,000 atmospheres, which turned graphite into diamond directly, without the aid of a catalyst. These are synthetic diamonds, which are not used to adorn the necks of duchesses, but for far more productive purposes — as cutting tools in industry.

Phase Transitions

A most important field of investigation concerns what are known as phase transitions — the critical point where matter changes from solid to liquid or from liquid to vapor; or the change from non-magnet to magnet; or from conductor to superconductor. All these processes are different, yet it has now been established beyond doubt that they are similar, so much so that the mathematics applied to one of these experiments can be applied to many others. This is a very clear example of a qualitative leap, as the following passage from James Gleick shows:

> Like so much of chaos itself, phase transitions involve a kind of macroscopic behavior that seems hard to predict by looking at the microscopic details. When a solid is heated, its molecules vibrate with the added energy. They push outward against their bonds and force the substance to expand. The more heat, the more expansion. Yet at a certain temperature and pressure, the change becomes sudden and discontinuous. A rope has been stretching; now it breaks. Crystalline form dissolves, and the molecules slide away from one another. They obey fluid laws that could not have been inferred from any aspect of the solid. The average atomic energy has barely changed, but the material — now a liquid, or a magnet, or a superconductor — has entered a new realm. [28]

Newton's dynamics were quite sufficient to explain large-scale phenomena but broke down for systems of atomic dimensions. Indeed, classical mechanics are still valid for most operations which do not involve very high speeds or the processes which take place at the subatomic level. Quantum mechanics will be dealt with in detail in another section. It represented a qualitative leap in science. Its relation to classical mechanics is similar to that between higher and lower mathematics and that between dialectics and formal logic. It can explain facts which classical mechanics could not, such as radioactive transformation, the transformation of matter into energy. It gave rise to new branches of science — theoretical chemistry, capable of solving previously insoluble problems. The theory of metallic magnetism underwent a fundamental change, making possible brilliant discoveries in the flow of electricity through metals. A whole series of theoretical difficulties were eliminated, once the new standpoint was accepted. But for a long time it met with a stubborn resistance, precisely because its results clashed head-on with the traditional mode of thinking and the laws of formal logic.

Modern physics furnishes a wealth of examples of the laws of dialectics, starting with quantity and quality. Take, for instance, the relation between the different kinds of electromagnetic wave and their frequencies, that is, the speed with which they pulsate. Maxwell's work, which Engels was very interested in, showed that electromagnetic waves and light waves were of the same kind. Quantum mechanics later showed that the situation is much more complex and contradictory, but at lower frequencies, the wave theory holds good.

The properties of different waves is determined by the number of oscillations per second. The difference is in the frequency of the waves, the speed with which they pulsate, the number of vibrations per second. That is to say, quantitative changes give rise to different kinds of wave signals. Translated into colors, red light indicates light waves of low frequency. An increased rate of vibration turns the color to orange-yellow, then to violet, then to the invisible ultra-violet and X-rays and finally to gamma rays. If we reverse the process, at the lower end, we go from infrared and heat rays to radio-waves. Thus, the same phenomenon manifests itself differently, in accordance with a higher or lower frequency. Quantity changes into quality.

The Electromagnetic Spectrum

The Electromagnetic Spectrum

Frequency in oscillations/sec	Name	Rough behavior
10^2	Electrical disturbance	Field
5×10^5-10^6	Radio Broadcast	
10^8	FM-TV	Waves
10^{10}	Radar	
5×10^{14}-10^{15}	Light	
10^{18}	X-rays	
10^{21}	γ-rays, nuclear	Particle
10^{24}	γ-rays, "artificial"	
10^{27}	γ-rays, in cosmic rays	

Source: R. P. Feynman, *Lectures on Physics*, chapter 2, p. 7, Table 2-1.

Order Out of Chaos

The law of quantity and quality also serves to shed light on one of the most controversial aspects of modern physics, the so-called "uncertainty principle," which we will examine in greater detail in another section. Whereas it is impossible to know the exact position and velocity of an individual subatomic particle, it is possible to predict with great accuracy the behavior of large numbers of particles. A further example: radioactive atoms decay in a way that makes a detailed prediction impossible. Yet large numbers of atoms decay at a rate so statistically reliable that they are used by scientists as natural "clocks" with which they calculate the age of the earth, the sun and the stars. The very fact that the laws governing the behavior of subatomic particles are different to those which function at the "normal" level is itself an example of the transformation of quantity into quality. The precise point at which the laws of the small-scale phenomena cease to apply was defined by the quantum of action laid down by Max Planck in 1900.

At a certain point, the concatenation of circumstances causes a qualitative leap whereby inorganic matter gives rise to organic matter. The difference between inorganic and organic matter is only relative. Modern science is well on the way to discovering exactly how the latter arises from the former. Life itself consists of atoms organized in a certain way. We are all a collection of atoms but not "merely" a collection of atoms. In the astonishingly complex arrangement of our genes, we have an infinite number of possibilities. The task of allowing each individual to develop these possibilities to the fullest extent is the real task of socialism.

Molecular biologists now know the complete DNA sequence of an organism, but cannot deduce from this how the organism assembles itself during its development, any more than knowledge of the structure of H_2O provides an understanding of the quality of liquidity. An analysis of the chemicals and cells of the body does not add up to a formula for life. The same is true of the mind itself. Neuroscientists have a great deal of data about what the brain does. The human brain consists of ten billion neurons, each of which has an average of a thousand links with other neurons. The fastest computer is capable of performing around a billion operations a second. The brain of a fly sitting on a wall carries out 100 billion operations in the same time. This comparison gives an idea of the vast difference between the human brain and even the most advanced computer.

The enormous complexity of the human brain is one of the reasons why idealists have attempted to surround the phenomenon of mind with a mystical aura. Knowledge of the details of individual neurons, axons and synapses, is not sufficient to explain the phenomenon of thought and emotion. However, there is nothing mystical about it. In the language of complexity theory, both mind and life are emergent phenomena. In the language of dialectics, the leap from quantity to quality means that the whole possesses qualities which cannot be deduced from the sum of the parts or reduced to it. None of the neurons is itself conscious. Yet the sum total of neurons and their connections are. Neural networks are non-linear dynamical systems. It is the complex activity and interactions between the neurons which produce the phenomenon we call consciousness.

The same kind of thing can be seen in large numbers of multi-component systems in the most varied spheres. Studies of ant colonies at Bath University have shown how behavior not witnessed in individual ants appears in a colony. A single ant, left to itself, will wander around at random, foraging and resting at irregular intervals. However, when the observation shifts to a whole colony of ants it immediately becomes evident that they become active at perfectly regular intervals. It is thought that this maximizes the effectiveness of their labors: if they all work together, one ant is unlikely to repeat a task just performed by another. The degree of coordination at the level of an ant colony is such that some people have thought of it as a single animal, rather than a colony. This too is a mystical presentation of a phenomenon which exists on many levels in nature and in animal and human society, and which can only be understood in terms of the dialectical relation between whole and part.

We can see the law of the transformation of quantity into quality at work when we consider the evolution of the species. In biological terms a specific "breed" or "race" of animal is defined by its capacity to inter-breed. But as evolutionary modifications take one group further away from another a point is reached where they can no longer inter-breed. At this point a new species has been formed. Paleontologists Stephen Jay Gould and Niles Eldredge have demonstrated that these processes are some times slow and protracted and at other times extremely rapid. Either way, they show how a gradual accumulation of small changes at a certain point provokes a qualitative change. Punctuated equilibria is the term used by these biologists to describe long periods of stability, interrupted by sudden bursts of change. When this idea was proposed by Gould and Eldredge of the American Museum to Natural History in 1972, it provoked an acrimonious debate among biologists, for whom, until then, Darwinian evolution was synonymous with gradualism.

For a long time, it was thought that evolution precluded such drastic changes. It was pictured as a slow, gradual change. However, the fossil record, although incomplete, presents a very different picture, with long periods of gradual evolution punctuated by violent upheavals, accompanied by the mass extinction of some species and the rapid rise of others. Whether or not the dinosaurs became extinct as a consequence of a meteorite colliding with the earth, it seems highly improbable that most of the great extinctions were caused in this way. While external phenomena, including meteorite or comet impacts, can play a role as "accidents" in the evolutionary process, it is necessary to seek an explanation of evolution as a result of its internal laws. The theory of "punctuated equilibria," which is now supported by most paleontologists, represents a decisive break with the old gradualist interpretation of Darwinism, and presents a truly dialectical picture of evolution, in which long periods of stasis are interrupted by sudden leaps and catastrophic changes of all kinds.

There is an endless number of examples of this law covering a very wide field. Is it possible now to continue to doubt the validity of this extremely important law ? Is it really justified to continue to ignore it or to write it off as a subjective invention which has been arbitrarily applied to diverse phenomena which bear no relation to one another? We see how in physics the study of phase transitions has led to the conclusion that apparently unrelated changes — of the boiling of liquids and the magnetizing of metals — all follow the same rules. It is only a matter of time before similar connections will be established which will reveal beyond a shadow of doubt that the law of the transformation of quantity into quality is indeed one of the most fundamental laws of nature.

Whole and Part

According to formal logic, the whole is equal to the sum of its parts. On closer examination, however, this is seen not to be true. In the case of living organisms it is manifestly not the case. A rabbit cut up in a laboratory, and reduced to its constituent parts is no longer a rabbit. This fact has been grasped by the advocates of chaos theory and complexity. Whereas classical physics, with its linear systems, accepted that the whole was precisely the sum of its parts, the non-linear logic of complexity maintains the opposite proposition, in complete agreement with dialectics:

"The whole is almost always equal to a great deal more than a sum of its parts," says Waldrop. "And the mathematical expression of that property — to the extent that such systems can be described by mathematics at all — is a non-linear equation: one whose graph is curvy." [29]

We have already quoted the examples of the qualitative changes in chemistry used by Engels in *Anti-Dühring*. While these examples remain valid, they by no means tell the whole story. Engels was limited, of course, by the scientific knowledge of his time. Today it is possible to go much further. The classical atomic theory of chemistry sets out from the idea that any combination of atoms into a greater unity can only be an aggregate of these atoms, that is, a purely quantitative relation. The union of atoms into molecules was seen as a simple juxtaposition. Chemical formulae such as H_2O, H_2SO_4, etc. presuppose that each of the atoms remains basically unchanged even when it enters a new combination to form a molecule.

This reflected precisely the mode of thinking of formal logic, which states that the whole is only the sum of the parts. Thus, since the molecular weight equals the sum of the weights of the respective atoms, it was assumed that the atoms themselves had remained unchanged, having entered into a purely quantitative relationship. However, many of the properties of the compounds could not be determined in this way. Indeed, most chemical properties of compounds differ considerably from those of the elements of which they are made up. The so-called "principle of juxtaposition" does not explain these changes. It is one-sided, inadequate and, in a word, wrong.

Modern atomic theory has shown the incorrectness of this idea. While accepting that complex structures must be explained in terms of aggregates of more elementary factors, it has shown that the relations between these elements are not merely indifferent and quantitative, but dynamic and dialectical. The elementary particles which make up the atoms interact constantly, passing into each other. They are not fixed constants but are at every moment both themselves and something else at the same time. It is precisely this dynamic relationship which gives the resulting molecules their particular nature, properties and specific identity.

In this new combination the atoms are and are not themselves. They combine in a dynamic way to produce an entirely different entity, a different relationship, which, in turn, determines the behavior of its component parts. Thus, we are not dealing merely with a lifeless "juxtaposition," a mechanical aggregate, but with a process. In order to understand the nature of an entity it is therefore entirely insufficient to reduce it to its individual atomic components. It is necessary to understand its dynamic interrelations, that is, to arrive at a dialectical, not a formal, analysis.

David Bohm was one of the few to provide a worked-out theoretical alternative to the subjectivist "Copenhagen interpretation" of quantum mechanics. Bohm's analysis, which is clearly influenced by the dialectical method, advocates a radical re-thinking of quantum mechanics and a new way of looking at the relationship between whole and parts. He points out that the usual interpretation of quantum theory does not give an adequate idea of just how far-reaching was the revolution affected by modern physics.

"Indeed," says Bohm, "when this interpretation is extended to field theories, not only the inter-relationships of the parts, but also their very existence is seen to flow out of the law of the whole. There is therefore nothing left of the classical scheme, in which the whole is derived from preexistent parts related in predetermined ways. Rather, what we have is reminiscent of the relationship of whole and parts in an organism, in which each organ grows and sustains itself in a way that depends crucially on the whole." [30]

A molecule of sugar can be broken down into its constituent parts of single atoms but then it is no longer sugar. A molecule cannot be reduced to its component parts without losing its identity. This is precisely the problem when we try to treat complex phenomena from a purely quantitative point of view. The resulting over-simplification leads to a distorted and one-sided picture of the natural world since the qualitative aspect is entirely left out of account. It is precisely through quality that we are able to distinguish one thing from another. Quality lies at the basis of all our knowledge of the world because it expresses the fundamental reality of all things, showing the critical boundaries that exist at all levels of material reality. The exact point at which small changes of degree give rise to a change of state is one of the most fundamental problems of science. It is a question which occupies a central place in dialectical materialism.

Complex Organisms

Life itself arises from a qualitative leap from inorganic to organic matter. The explanation of the processes by which this occurred constitutes one of the most important and exciting problems of present-day science. The advances of chemistry, analyzing in great detail the structures of complex molecules, predicting their behavior with great accuracy and identifying the role of particular molecules in living systems, paved the way for the emergence of new sciences, biochemistry and biophysics, dealing respectively with the chemical reactions that take place in living organisms and the physical phenomena involved in living processes. These, in turn, have been merged together in molecular biology, which has registered the most amazing advances in recent years.

In this way, the old fixed divisions separating organic and inorganic matter have been entirely abolished. The early chemists drew a rigid distinction between the two. Gradually, it was understood that the same chemical laws applied to organic as to inorganic molecules. All substances containing carbon (with the possible exception of a few simple compounds like carbon dioxide) are characterized as organic. The rest are inorganic. Only carbon atoms are capable of forming very long chains, thus giving rise to the possibility of an infinite variety of complex molecules.

In the 19th century chemists analyzed the properties of "albuminous" substances (from the Latin word for egg-white). From this, it was discovered that life was dependent upon proteins, large molecules made up of amino acids. At the beginning of the 20th century, when Planck was making his breakthrough in physics, Emil Fischer was attempting to join up amino-acids in chains in such a manner that the carboxyl group of one amino-acid was always linked to the amino group of the next. By 1907, he had succeeded in synthesizing a chain of eighteen amino-acids. Fischer called these chains peptides, from the Greek word "to digest," because he thought that proteins would break down into such chains in the process of digestion. This theory was finally proven by Max Bergmann in 1932.

These chains were still too simple to produce the complex polypeptide chains needed to create proteins. Moreover, the task of deciphering the structure of a protein molecule itself was incredibly difficult. The properties of each protein depends on its exact relation to each amino acid on the molecular chain. Here too, quantity determines quality. This posed a seemingly insurmountable problem for biochemists, since the number of possible arrangements in which nineteen amino acids can appear on a chain comes to nearly 120 million billion. A protein the size of serum albumen, made up of more than 500 amino acids, therefore has a number of possible arrangements of about 10600, that is, 1 followed by 600 zeros. The complete structure of a key protein molecule — insulin — was established for the first time by the British biochemist Fredrich Sanger in 1953. Using the same method, other scientists succeeded in deciphering the structure of a whole series of other protein molecules. Later, they succeeded in synthesizing protein in the laboratory. It is now possible to synthesize many proteins, including one as complex as the human growth hormone which involves a chain of 188 amino acids.

Life is a complex system of interactions, involving an immense number of chemical reactions which proceed continuously and rapidly. Every reaction in the heart, blood, nervous system, bones and brain interacts with every other part of the body. The workings of the simplest living body are far more complicated than the most advanced computer, permitting rapid movement, swift reactions to the slightest change in the environment, constant adjustments to changing conditions, internal and external. Here, most emphatically, the whole is more than the sum of the parts. Every part of the body, every muscular and nervous reaction, depends upon all the rest. Here we have a dynamic and complex, in other words, dialectical, interrelationship which alone is capable of creating and sustaining the phenomenon we know as life.

The process of metabolism means that, at any given moment, the living organism is constantly changing, taking in oxygen, water, and food (carbohydrates, fats, proteins, minerals and other raw materials), negating these by transforming them into the materials needed to sustain and develop life and excreting waste products. The dialectical relationship between whole and part manifests itself in the different levels of complexity in nature, reflected in the different branches of science.

a) Atomic interactions and the laws of chemistry determine the laws of biochemistry, but life itself is qualitatively different.

b) The laws of biochemistry "explain" all the processes of human interaction with the environment. And yet human activity and thought are qualitatively different to the biological processes that constitute them.

c) Each individual person, in turn, is a product of his or her physical and environmental development. Yet the complex interactions of the sum total of individuals which make up a society are also qualitatively different. In each of these cases the whole is greater than the sum of the parts and obeys different laws.

In the last analysis, all human existence and activity is based on the laws of motion of atoms. We are part of a material universe, which is a continuous whole, functioning according to its inherent laws. And yet, when we pass from a) to c), we make a series of qualitative leaps, and must operate with different laws at different "levels"; c) is based upon b) and b) is based upon a), but nobody in their right mind would seek to explain the complex movements in human society in terms of atomic forces. For the same reason, it is absolutely futile to reduce the problem of crime to the laws of genetics.

An army is not merely the sum total of individual soldiers. The very act of combining in a massive force, organized on military lines transforms the individual soldier both physically and morally. As long as the cohesiveness of the army is maintained, it represents a formidable force. This is not only a question of numbers. Napoleon was well aware of the importance of morale in war. As part of a disciplined numerous fighting force, the individual soldier is capable of achieving feats of bravery and self-sacrifice in situations of extreme danger, of which, under normal conditions, as an isolated individual, he would never imagine himself capable. Yet he remains the same person as before. The moment the cohesiveness of the army breaks down under the impact of defeat, the whole dissolves into its individual "atoms," and the army becomes a demoralized rabble.

Engels was very interested in military tactics, for which Marx's daughters nicknamed him "the General." He closely followed the progress of the American Civil War and the Crimean War, about which he wrote many articles. In *Anti-Dühring*, he shows how the law of quantity and quality relates to military tactics, for example, in the relative fighting capacity of the highly disciplined soldiers of Napoleon and the Egyptian (Mameluke) cavalry:

"In conclusion, we shall call one more witness for the transformation of quantity into quality, namely Napoleon. He describes the combat between the French cavalry, who were bad riders but disciplined, and the Mamelukes, who were undoubtedly the best horsemen of their time for single combat but who lacked discipline, as follows:

"'Two Mamelukes were undoubtedly more than a match for three Frenchmen; 100 Mamelukes were equal to 100 Frenchmen; 300 Frenchmen could generally beat 300 Mamelukes, and 1,000 Frenchmen invariably defeated 1,500 Mamelukes.' Just as with Marx a definite, though varying, minimum sum of exchange-value was necessary to make possible its transformation into capital, so with Napoleon a detachment of cavalry had to be of a definite minimum number in order to permit the force of discipline, embodied in close order and planned utilization, to manifest itself and even rise superior to greater numbers of irregular cavalry, who were better mounted, more dexterous horsemen and fighters, and at least as brave as the former." [31]

The Molecular Process of Revolution

The process of chemical reaction involves crossing a decisive barrier known as a transition state. At this point, before the reactants become products, they are neither one thing nor the other. Some of the old bonds are breaking and other new ones are being formed. The energy needed to pass this critical point is known as Gibbs energy. Before a molecule can react, it requires a quantity of energy which, at a certain point, brings it to the transition state. At normal temperatures only a minute fraction of the reactant molecules possess sufficient energy. At a greater temperature, a higher proportion of the molecules will have this energy. That is why heating is one way to speed up a chemical reaction. The process can be assisted by the use of catalysts, which are widely used in industry. Without catalysts, many processes, though they would still take place, would be so slow that they would be uneconomic. The catalyst cannot change the composition of the substances involved nor can it alter the Gibbs energy of the reactants, but it can provide an easier pathway between them.

There are certain analogies between this phenomenon and the role of the individual in history. It is a common misconception that Marxism has no place for the role of individuals in shaping their own destiny. According to this caricature, the materialist conception of history reduces everything to "the productive forces." Human beings are seen as mere blind agents of economic forces or marionettes dancing on the strings of historical inevitability. This mechanistic view of the historic process (economic determinism) has nothing in common with the dialectical philosophy of Marxism.

Historical materialism sets out from the elementary proposition that men and women make their own history. But, contrary to the idealist notion of human beings as absolutely free agents, Marxism explains that they are limited by the actual material conditions of the society into which they are born. These conditions are shaped in a fundamental way by the level of development of the productive forces, which is the ultimate ground upon which all human culture, politics and religion, rest. However, these things are not directly shaped by economic development but can and do take on a life of their own. The extremely complex relation between all these factors has a dialectical character, not a mechanical one. Individuals do not choose the conditions into which they are born. They are "given." Nor is it possible, as idealists imagine, for individuals to impose their will upon society, simply because of the greatness of their intellect or the strength of their character. The theory that history is made by "great individuals" is a fairy-story fit to amuse five-year-olds. It has approximately the same scientific value as the "conspiracy theory" of history, which attributes revolutions to the malign influence of "agitators."

Every worker knows that strikes are not caused by agitators but by bad wages and conditions. Contrary to the impression sometimes given by certain sensationalist newspapers, strikes are not common occurrences. For many years, a factory or workplace can remain apparently peaceful. The workforce may not react, even when their wages and conditions are attacked. This is especially true in conditions of mass unemployment or when there is no lead from the tops of the trade unions. This apparent indifference of the majority often leads the minority of activists to despair. They draw the mistaken conclusion that the rest of the workers are "backward," and will never do anything. But, in fact, beneath the surface of apparent tranquility, changes are taking place. A thousand small incidents, pin-pricks, injustices, injuries, gradually leave their mark on the consciousness of the workers. This process was aptly described by Trotsky as "the molecular process of revolution." It is the equivalent of the Gibbs energy in a chemical reaction.

In real life, as in chemistry, molecular processes take their time. No chemist would ever complain because the anticipated reaction was taking a long time, especially if the conditions for a speedy reaction (high temperature, etc.) were absent. But eventually, the chemical transition state is reached. At this point, the presence of a catalyst is of great assistance in bringing the process to a successful conclusion, in the speediest and most economical manner. In the same way, at a given point, the accumulated mood of discontent in the workplace boils over. The whole situation is transformed in the space of 24 hours. If the activists are not prepared, if they have allowed themselves to be deceived by the surface calm of the previous period, they will be taken completely off guard.

In dialectics, sooner or later, things change into their opposite. In the words of the Bible, "the first shall be last and the last shall be first." We have seen this many times, not least in the history of great revolutions. Formerly backward and inert layers can catch up with a bang. Consciousness develops in sudden leaps. This can be seen in any strike. And in any strike we can see the elements of a revolution in an undeveloped, embryonic form. In such situations, the presence of a conscious and audacious minority can play a role quite similar to that of a catalyst in a chemical reaction. In certain instances, even a single individual can play an absolutely decisive role.

In November 1917, the fate of the Russian Revolution was ultimately determined by two men - Lenin and Trotsky. Without them, there is no doubt that the revolution would have been defeated. The other leaders, Kamenev, Zinoviev and Stalin, came under the pressure of other classes and capitulated. The question here is not one of abstract "historical forces" but the concrete one of the degree of preparation, foresight, personal courage and ability of leaders.

After all, we are talking about a struggle of living forces not a simple mathematical equation.

Does this mean then that the idealist interpretation of history is correct? Is it all decided by great individuals? Let the facts speak for themselves. For a quarter of a century before 1917, Lenin and Trotsky had spent most of their lives more or less isolated from the masses, often working with very small groups of people. Why were they unable to have the same decisive effect, for example, in 1916? Or in 1890? Because the objective conditions were absent. In the same way, a union activist who continually called for a strike when there was no mood for action, would soon end up a laughing-stock. Similarly, when the revolution was isolated in conditions of unspeakable backwardness and the class balance of forces changed, neither Lenin nor Trotsky could prevent the rise of the bureau-cratic counterrevolution headed by a man in every way their inferior, Stalin. Here, in a nutshell, we have the dialectical relation between the subjective and objective factor in human history.

The Unity and Interpenetration of Opposites

Everywhere we look in nature, we see the dynamic co-existence of opposing tendencies. This creative tension is what gives life and motion. That was already understood by Heraclitus (c. 500 B.C.) two and a half thousand years ago. It is even present in embryo in certain Oriental religions, as in the idea of the ying and yang in China, and in Buddhism. Dialectics appears here in a mystified form, which nonetheless reflects an intuition of the workings of nature. The Hindu religion contains the germ of a dialectical idea, when it poses the three phases of creation (Brahma), maintenance or order (Vishnu) and destruction or disorder (Shiva). In his interesting book on the mathematics of chaos, Ian Stewart points out that the difference between the gods Shiva, "the Untamed," and Vishnu is not the antagonism between good and evil, but that the two principles of harmony and discord together underlie the whole of existence.

"In the same way," he writes, "mathematicians are beginning to view order and chaos as two distinct manifestations of an underlying determinism. And neither exists in isolation. The typical system can exist in a variety of states, some ordered, some chaotic. Instead of two opposed polarities, there is a continuous spectrum. As harmony and discord combine in musical beauty, so order and chaos combine in mathematical beauty." [32]

In Heraclitus, all this was in the nature of an inspired guess. Now this hypothesis has been confirmed by a huge amount of examples. The unity of opposites lies at the heart of the atom, and the entire universe is made up of molecules, atoms, and subatomic particles. The matter was very well put by R. P.

Feynman: "All things, even ourselves, are made of fine-grained, enormously strongly interacting plus and minus parts, all neatly balanced out." [33]

The question is: how does it happen that a plus and a minus are "neatly balanced out?" This is a contradictory idea! In elementary mathematics, a plus and a minus do not "balance out." They negate each other. Modern physics has uncovered the tremendous forces which lie at the heart of the atom. Why do the contradictory forces of electrons and protons not cancel each other out? Why do atoms not merely fly apart? The current explanation refers to the "strong force" which holds the atom together. But the fact remains that the unity of opposites lies at the basis of all reality.

Within the nucleus of an atom, there are two opposing forces, attraction and repulsion. On the one hand, there are electrical repulsions which, if unrestrained, would violently tear the nucleus apart. On the other hand, there are powerful forces of attraction which bind the nuclear particles to each other. This force of attraction, however, has its limits, beyond which it is unable to hold things together. The forces of attraction, unlike repulsion, have a very short reach. In a small nucleus they can keep the forces of disruption in check. But in a large nucleus, the forces of repulsion cannot be easily dominated.

Beyond a certain critical point, the bond is broken and a qualitative leap occurs. Like an enlarged drop of water, it is on the verge of breaking apart. When an extra neutron is added to the nucleus, the disruptive tendency increases rapidly. The nucleus breaks up, forming two smaller nuclei, which fly apart violently, releasing a vast amount of energy. This is what occurs in nuclear fission. However, analogous processes may be seen at many different levels of nature. Water falling on a polished surface will break up into a complex pattern of droplets. This is because two opposing forces are at work: gravity, which tries to spread out the water in a flat film spread over the whole surface, and surface tension, the attraction of one water molecule to another, which tries to pull the liquid together, forming compact globules.

Nature seems to work in pairs. We have the "strong" and the "weak" forces at the subatomic level; attraction and repulsion; north and south in magnetism; positive and negative in electricity; matter and anti-matter; male and female in biology; odd and even in mathematics; even the concept of "left and right handedness" in relation to the spin of subatomic particles. There is a certain symmetry, in which contradictory tendencies, to quote Feynman, "balance themselves out," or, to use the more poetical expression of Heraclitus, "agree with each other by differing like the opposing tensions of the strings and bow of a musical instrument." There are two kinds of matter, which can be called positive and negative. Like kinds repel and unlike attract.

Positive and Negative

Positive is meaningless without negative. They are necessarily inseparable. Hegel long ago explained that "pure being" (devoid of all contradiction) is the same as pure nothing, that is, an empty abstraction. In the same way, if everything were white, it would be the same for us as if everything were black. Everything in the real world contains positive and negative, being and not being, because everything is in a state of constant movement and change. Incidentally, mathematics shows that zero itself is not equal to nothing.

"Zero," writes Engels, "because it is the negation of any definite quantity, is not therefore devoid of content. On the contrary, zero has a very definite content. As the border-line between all positive and negative magnitudes, as the sole really neutral number, which can be neither positive nor negative, it is not only a very definite number, but also in itself more important than all other numbers bounded by it. In fact, zero is richer in content than any other number. Put on the right of any other number, it gives to the latter, in our system of numbers, the tenfold value. Instead of zero one could use here any other sign, but only on the condition that this sign taken by itself signifies zero = 0. Hence it is part of the nature of zero itself that it finds this application and that it alone can be applied in this way. Zero annihilates every other number with which it is multiplied; united with any other number as divisor or dividend, in the former case it makes this infinitely large, in the latter infinitely small; it is the only number that stands in a relation of infinity to every other number. 0/0 can express every number between $-$ and $+$, and in each case represents a real magnitude." [34]

The negative magnitudes of algebra only have meaning in relation to the positive magnitudes, without which they have no reality whatsoever. In the differential calculus, the dialectical relation between being and not being is particularly clear. This was extensively dealt with by Hegel in his *Science of Logic*. He was greatly amused by the perplexity of the traditional mathematicians, who were shocked by the use of a method which makes use of the infinitesimally small, and "cannot do without the suggestion that a certain quantity is not equal to nil but is so inconsiderable that it may be neglected," and yet always obtains an exact result. [35]

Moreover, everything is in a permanent relation with other things. Even over vast distances, we are affected by light, radiation, gravity. Undetected by our senses, there is a process of interaction, which causes a continual series of changes. Ultra-violet light is able to "evaporate" electrons from metal surfaces in much the same way as the sun's rays evaporate water from the ocean. Banesh Hoffmann states: "It is still a strange and awe-inspiring thought, that you and I are thus rhythmically exchanging particles with one another, and with the earth and the beasts of the earth, and the sun and the moon and the stars, to the uttermost galaxy." [36]

The Dirac equation for the energy of an individual electron involves two answers — one positive and one negative. It is similar to the square root of a number, which can either be positive or negative. Here, however, the negative answer implies a contradictory idea — negative energy. This appears to be an absurd concept from the standpoint of formal logic. Since energy and mass are equivalent, negative energy, in turn, means negative mass. Dirac himself was disturbed by the implications of his own theory. He was compelled to predict the existence of particles which would be identical to the electron, but with a positive electric charge, a previously unheard of matter.

On August 2nd, 1932, Robert Millikan and Carl D. Anderson of the California Institute of Technology discovered a particle the mass of which was clearly that of an electron, but moving in the opposite direction. This was not an electron, proton or neutron. Anderson described it as a "positive electron" or positron. This was a new kind of matter — antimatter — predicted by Dirac's equations. Subsequently, it was discovered that electrons and positrons, when they meet, annihilate each other, producing two photons (two flashes of light). In the same way, a photon passing through matter could split to form a virtual electron and a positron.

The phenomenon of oppositeness exists in physics, where, for example, every particle has its anti-particle (electron and positron, proton and anti-proton, etc.). These are not merely different, but opposites in the most literal sense of the word, being identical in every respect, except one: they have opposite electrical charges — positive and negative. Incidentally, it is a matter of indifference which one is characterized as negative and which positive. The important thing is the relationship between them.

Every particle possesses the quality known as spin, expressed as a plus or a minus, depending on its direction. Strange as it may seem, the opposite phenomena of left and right handedness, which is known to play a fundamental role in biology, also has its equivalent at the subatomic level. Particles and waves stand in contradiction to each other. The Danish physicist Niels Bohr referred to them, rather confusingly, as "complementary concepts," by which he meant precisely that they exclude one another.

The most recent investigations of particle physics are casting light on the deepest level of matter so far discovered — quarks. These particles also have opposing "qualities" which are not comparable to ordinary forms, so physicists are obliged to make up new, artificial qualities to describe them. Thus we have up-quarks, down-quarks, charm-quarks, strange quarks, and so on. Although the qualities of quarks have still to be thoroughly explored, one thing is clear:

that the property of oppositeness exists at the most fundamental levels of matter yet known to science.

This universal phenomenon of the unity of opposites is, in reality, the motor-force of all motion and development in nature. It is the reason why it is not necessary to introduce the concept of external impulse to explain movement and change — the fundamental weakness of all mechanistic theories. Movement, which itself involves a contradiction, is only possible as a result of the conflicting tendencies and inner tensions which lie at the heart of all forms of matter.

The opposing tendencies can exist in a state of uneasy equilibrium for long periods of time, until some change, even a small quantitative change, destroys the equilibrium and gives rise to a critical state which can produce a qualitative transformation. In 1936, Bohr compared the structure of the nucleus to a drop of liquid, for example, a raindrop hanging from a leaf. Here the force of gravity struggles with that of surface tension striving to keep the water molecules together. The addition of just a few more molecules to the liquid renders it unstable. The enlarged droplet begins to shudder, the surface tension is no longer able to hold the mass to the leaf and the whole thing falls.

Nuclear Fission

This apparently simple example, of which many equivalents can be observed a hundred times in daily experience, is a fairly close analogy to the processes at work in nuclear fission. The nucleus itself is not at rest, but in a constant state of change. In one quadrillionth of a second, there have already been billions of random collisions of particles. Particles are constantly entering and leaving the nucleus. Nevertheless, the nucleus is held together by what is often described as the strong force. It remains in a state of unstable equilibrium, "on the edge of chaos," as chaos theory would put it.

As in a drop of liquid which quivers as the molecules move around inside it, the particles are constantly moving, transforming themselves, exchanging energy. Like an enlarged raindrop, the bond between the particles in a large nucleus is less stable, and more likely to break up. The steady release of alpha particles from the surface of the nucleus makes it smaller and steadier. As a result, it may become stable. But it was discovered that by bombarding a large nucleus with neutrons they can be made to break up, releasing part of the colossal amounts of energy locked up in the atom. This is the process of nuclear fission. This process can occur even without the introduction of particles from without. The process of spontaneous fission (radio active decay) is going on all the time in nature. In one second, a pound of uranium experiences four spontaneous fissions, and alpha particles are emitted from around eight million nuclei. The heavier the nucleus, the more likely the process of fission becomes.

The unity of opposites lies at the root of life itself. When spermatozoa were first discovered, they were believed to be "homunculae," perfectly formed miniature human beings, which — like Flopsy in *Uncle Tom's Cabin* —"just grow'd." In reality, the process is far more complex and dialectical. Sexual reproduction depends on the combination of a single sperm and egg, in a process in which both are destroyed and preserved at the same time, passing on all the genetic information necessary for the creation of an embryo. After undergoing a whole series of transformations, bearing a striking resemblance to the evolution of all life from the division of a single cell, eventually results in an entirely new individual. Moreover, the result of this union contains the genes of both parents, but in such a way as to be different from either. So what we have is not simple reproduction, but a real development. The increased diversity made possible by this is one of the great evolutionary advantages of sexual reproduction.

The laws of formal logic have received a humiliating drubbing in the field of modern physics, where they have shown themselves to be hopelessly inadequate to deal with the contradictory processes that occur at the subatomic level. Particles which disintegrate so rapidly that it is difficult to say whether they exist or not, pose insurmountable problems for a system which attempts to ban all contradiction from nature and thought. This immediately leads to new contradictions of an insoluble character. Thought finds itself in opposition to the facts established and repeatedly confirmed by experiment and observation. The unity of the proton and the electron is a neutron. But if a positron should unite with a neutron, the result would be the shedding of an electron and the neutron would change into a proton. By means of this ceaseless process, the universe makes and re-makes itself over and over again. No need then for any external force, no "first impulse," as in classical physics. No need for anything whatsoever, except the infinite, restless movement of matter in accordance with its own objective laws.

Contradictions are found at all levels of nature, and woe betide the logic that denies it. Not only can an electron be in two or more places at the same time, but it can move simultaneously in different directions. We are sadly left with no alternative but to agree with Hegel: they are and are not. Things change into their opposite. Negatively-charged electrons become transformed into positively-charged positrons. An electron that unites with a proton is not destroyed, as one might expect, but produces a new particle, a neutron, with a neutral charge.

Polar Opposites?

Polarity is an all-pervasive feature in nature. It does not only exist as the North and South poles of the earth. Polarity is to be found in the sun and moon and other planets. It also exists at the subatomic level, where nuclei behave precisely as if they possess not one but two pairs of magnetic poles.

"Dialectics," wrote Engels, "has proved from the result of our experience of nature so far that all polar opposites in general are determined by the mutual action of the two opposite poles on each other, that the separation and opposition of these poles exist only within their mutual connection and union, and, conversely, that their union exists only in their separation and their mutual connection only in their opposition. This once established, there can be no question of a final canceling out of repulsion and attraction, or of a final partition between the one form of motion in one half of matter and the other form in the other half, consequently there can be no question of mutual penetration or of absolute separation of the two poles. It would be equivalent to demanding in the first case that the north and south poles of a magnet should mutually cancel themselves out or, in the second case, that dividing a magnet in the middle between the two poles should produce on one side a north half without a south pole, and on the other side a south half without a north pole." [37]

There are some things which people consider to be absolute and immutable opposites. For instance, when we wish to convey the notion of extreme incompatibility, we use the term "polar opposites" — north and south are taken to be absolutely fixed and opposed phenomena. For over a thousand years, sailors have placed their faith in the compass, which guided them through unknown oceans, always pointing to this mysterious thing called the north pole. Yet closer analysis shows that the north pole is neither fixed nor stable. The earth is surrounded by a strong magnetic field (a geocentric axis dipole), as if a gigantic magnet were present at the center of the earth, aligned parallel to the earth's axis. This is related to the metallic composition of the earth's core, which is mainly made up of iron. In the 4.6 billion years since the solar system was formed, the rocks on earth have formed and reformed many times. And not only the rocks but everything else. Detailed measurements and investigation has now proved beyond doubt that the location of the magnetic poles is continually shifting. At the present time, they are moving very slowly — 0.3 degrees every million years. This phenomenon is a reflection of complex changes taking place in the earth, the atmosphere and the sun's magnetic field.

So small is the shift that for centuries it remained undetected. However, even this apparently imperceptible process of change gives rise to a sudden and spectacular leap, in which north becomes south and south becomes north. The changes in the location of the poles are accompanied by fluctuations in the strength of the magnetic field itself. This gradual process, characterized by a weakening of the magnetic field, culminates in a sudden leap. They change place, literally turning into their opposite. After this, the field starts to recover and gather strength again.

This revolutionary change has occurred many times during the history of the earth. It has been estimated that more than 200 such polar reverses have taken place in the last 65 million years; at least four have occurred in the last four million years. About 700,000 years ago, the north magnetic pole was located somewhere in Antarctica, the present south geographical pole. At this moment, we are in a process of weakening of the earth's magnetic field, which will inevitably culminate in a new reversal. The study of the earth's magnetic history is the special field of an entirely new branch of science — paleomagnetism — which is attempting to construct maps of all the reversals of the poles throughout the history of our planet. The discoveries of paleomagnetism, in turn, have provided conclusive evidence for the correctness of the theory of continental drift. When rocks (especially volcanic rocks) create iron-rich minerals, these respond to the earth's magnetic field as it exists at that moment, in the same way that pieces of iron react to a magnet, their atoms orienting in line with the field axis. In effect, they behave like a compass. By comparing the orientations of minerals in rocks of the same age in different continents, it is possible to trace the movements of the continents, including those which no longer exist, or only exist as tiny remnants.

In the reversal of the poles we see a most graphic example of the dialectical law of the unity and interpenetration of opposites. North and south — polar opposites in the most literal sense of these words — are not only inseparably united but determine each other by means of a complex and dynamic process, which culminates in a sudden leap in which supposedly fixed and immutable phenomena change into their opposites. And this dialectical process is not the arbitrary and fanciful invention of Hegel or Engels, but is conclusively demonstrated by the most recent discoveries of paleomagnetism. Truly it has been said, "when men are silent, the stones cry out!"

Attraction and Repulsion

This is an extension of the law of the unity and interpenetration of opposites. It is a law which permeates the whole of nature, from the smallest phenomena to the largest. At the base of the atom are immense forces of attraction and repulsion. The hydrogen atom, for example, is made up of a proton and an electron held together by electrical attraction. The charge carried by a particle may be positive or negative. Similar charges repel each other, whereas opposite kinds attract. Thus, within the nucleus, protons repel each other, but the nucleus is held together by tremendous nuclear force. In very heavy nuclei, however, the force of electrical repulsion can reach a point where the nuclear force is overcome and the nucleus flies apart.

Engels points out the universal role of attraction and repulsion:

> All motion consists in the interplay of attraction and repulsion. Motion, however, is only possible when each individual attraction is compensated by a corresponding repulsion somewhere else. Otherwise in time one side would get the preponderance over the other and then motion would finally cease. Hence all attractions and all repulsions in the universe must mutually balance one another. Thus the law of the indestructibility and uncreatability of motion is expressed in the form that each movement of attraction in the universe must have as its complement an equivalent movement of repulsion and vice versa; or, as ancient philosophy — long before the natural-scientific formulation of the law of conservation of force or energy — expressed it: the sum of all attractions in the universe is equal to the sum of all repulsions.

In Engels' day, the prevailing idea of motion was derived from classical mechanics, where motion is imparted from an external force which overcomes the force of inertia. Engels was quite scathing about the very expression "force," which he considered one-sided and insufficient to describe the real processes of nature. "All natural processes," he wrote, "are two-sided, they are based on the relation of at least two operative parts, action and reaction. The notion of force, however, owing to its origin from the action of the human organism on the external world, and further from terrestrial mechanics, implies that only one part is active, operative, the other part being passive, receptive." [38]

Engels was far in advance of his time in being highly critical of this notion, which had already been attacked by Hegel. In his *History of Philosophy*, Hegel remarks that "It is better (to say) that a magnet has a soul (as Thales expresses it) than that it has an attractive force; force is a kind of property that, separate from matter, is put forward as a kind of predicate — while soul, on the other hand, is this movement itself, identical with the nature of matter." This remark of Hegel, approvingly quoted by Engels, contains a profound idea — that motion and energy are inherent to matter. Matter is self-moving and self-organizing.

Even the word "energy" was not, in Engels' opinion, entirely adequate, although greatly to be preferred to "force." His objection was that "It still makes it appear as if 'energy' was something external to matter, something implanted in it. But in all circumstances it is to be preferred to the expression 'force.'" [39] The real relation has been demonstrated by Einstein's theory of the equivalence of mass and energy, which shows that matter and energy are one and the same thing. This was precisely the standpoint of dialectical materialism, as expressed by Engels, and even anticipated by Hegel, as the above quotation shows.

Every science has its own vocabulary, the terms of which frequently do not coincide with everyday usage. This can lead to difficulties and misunderstandings. The word "negation" is commonly understood to signify simple destruction, or annihilation. It is important to understand that in dialectics negation has an entirely different content. It means to negate and to preserve at the same time. One can negate a seed by crushing it underfoot. The seed is "negated" but not in the dialectical sense! If, however, the same seed is left to itself, under favorable conditions, it will germinate. It has thus negated itself as a seed and develops into a plant which, at a later stage, will die, producing new seeds.

Apparently, this represents a return to the starting point. However, as professional gardeners know, identical seeds vary from generation to generation giving rise to new species. Gardeners also know that certain strains can be artificially induced by selective breeding. It was precisely this artificial selection which gave Darwin a vital clue to the process of natural selection which takes place spontaneously throughout nature, and is the key to understanding the development of all plants and animals. What we have is not only change but actual development, generally proceeding from simpler to more complex forms, including the complex molecules of life itself, which, at a certain stage, arises from inorganic matter.

Consider the following example of negation from quantum mechanics. What occurs when an electron unites with a photon? The electron experiences a "quantum leap" and the photon disappears. The result is not some kind of mechanical unity or compound. It is the same electron as before but in a new state of energy. The same is true when the electron unites with a proton. The electron vanishes and there is a leap in the proton's state of energy and charge. The proton is the same as before but in a new state of energy and charge. It is now electrically neutral and becomes a neutron. Dialectically speaking, the electron has been negated and preserved at the same time. It has disappeared, but is not annihilated. It enters into the new particle and expresses itself as a change of energy and charge.

The ancient Greeks were well acquainted with the dialectic of discussion. In a properly conducted debate, an idea is put forward (the Thesis) and is then countered by the opposing view (the Antithesis) which negates it. Finally, through a thorough process of discussion, which explores the issue concerned from all points of view and discloses all the hidden contradictions, we arrive at a conclusion (the Synthesis). We may or may not arrive at agreement but by the very process of discussion, we have deepened our knowledge and understanding and raised the whole discussion onto a different plane.

It is quite evident that almost none of the critics of Marxism have taken the trouble to read Marx and Engels. It is frequently supposed, for example, that dialectics consists of "Thesis-Antithesis-Synthesis," which Marx is alleged to have copied from Hegel (who, in turn, was supposed to have copied it from the Holy Trinity) and applied to society. This childish caricature is still repeated by supposedly intelligent people today. As a matter of fact, not only is Marx's dialectical materialism the opposite of Hegel's idealist dialectic, but the dialectic of Hegel is itself very different from that of classical Greek philosophy.

Plekhanov rightly ridiculed the attempt to reduce the imposing edifice of Hegelian dialectic to the "wooden Triad" of Thesis-Antithesis-Synthesis. The advanced dialectics of Hegel bears approximately the same relation to that of the Greeks as modern chemistry to the primitive investigations of the alchemists. It is quite correct that the latter prepared the ground for the former, but to assert that they are "basically the same" is simply ludicrous. Hegel returned to Heraclitus, but on a qualitatively higher level, enriched by 2,500 years of philosophical and scientific advances. The development of dialectics is itself a dialectical process. Nowadays the word "alchemy" is used as a synonym for quackery. It conjures up all kinds of images of spells and black magic. Such elements were not absent from the history of alchemy, but its activities were by no means limited to this. In the history of science, alchemy played a most important role. Alchemy is an Arabic word, used for any science of materials. Charlatans there were, but not a few good scientists too! And chemistry is the Western word for the same thing. Many chemical words are, in fact, Arab in origin — acid, alkali, alcohol, and so on.

The alchemists set out from the proposition that it was possible to transmute one element into another. They tried for centuries to discover the "philosopher's stone," which they believed would enable them to turn base metal (lead) into gold. Had they succeeded, it would not have done them a lot of good, since the value of gold would have quickly sunk to that of lead! But that is another story. Given the actual level of technique at that time, the alchemists were attempting the impossible. In the end, they were forced to come to the conclusion that the transmutation of the elements was impossible. However, the endeavors of the alchemists were not in vain. In their pursuit of an unscientific hypothesis, the philosopher's stone, they actually did valuable pioneering work, developing the art of experiment, inventing equipment still used in laboratories today and describing and analyzing a wide range of chemical reactions. In this way, alchemy prepared the ground for the development of chemistry.

Modern chemistry was able to progress only by repudiating the alchemists' basic hypothesis — the transmutation of the elements. From the late 18th century onwards, chemistry developed on a scientific basis. By setting aside the grandiose aims of the past, it made giant steps forward. Then, in 1919, the English scientist Rutherford carried out an experiment involving the bombardment of nitrogen nuclei with alpha particles. This led to the breaching of the atomic nucleus for the first time. In so doing, he succeeded in transmuting one element (nitrogen) into another element (oxygen). The age-old quest of the alchemists had been resolved but not at all in a way they could have foreseen!

Now look at this process a bit more closely. We start with the thesis: a) the transmutation of the elements; this is then negated by its antithesis b) impossibility of transmuting the elements; this, in turn, is overturned by a second negation c) the transmutation of the elements. Here we must note three things. Firstly, each negation marks a definite advance, indeed, a qualitative leap forward. Secondly, each successive advance both negates the earlier stage, reacts against it, whilst preserving all that is useful and necessary in it. Lastly, the final stage — the negation of the negation — does not at all signify a return to the original idea (in this case, alchemy), but the reappearance of earlier forms on a qualitatively higher level. Incidentally, it is now possible to convert lead into gold, but would be too expensive to be worth the trouble!

Dialectics envisages the fundamental processes at work in the universe, in society and in the history of ideas, not as a closed circle, where the same processes merely repeat themselves in an endless mechanical cycle, but as a kind of open-ended spiral of development in which nothing is ever repeated exactly in the same way. This process can be clearly seen in the history of philosophy and science. The entire history of thought consists of an endless process of development through contradiction.

A theory is put forward which explains certain phenomena. This gradually gains acceptance, both through the accumulation of evidence which bears it out, and because of the absence of a satisfactory alternative. At a certain point, discrepancies appear, which are initially shrugged off as unimportant exceptions. Then a new theory emerges which contradicts the old one and seems to explain the observed facts better. Eventually, after a struggle, the new theory overthrows the existing orthodoxy. But new questions arise from this which in turn have to be resolved. Frequently, it appears that we return again to ideas which were earlier thought to be discredited. But this does not mean a return to the starting point. What we have is a dialectical process, involving a deeper and deeper understanding of the workings of nature, society, and ourselves. This is the dialectic of the history of philosophy and science.

Joseph Dietzgen, a companion of Marx and Engels, once said that an old man who looks back on his life may see it as an endless series of mistakes which, if he could only have his time back again, he would doubtless choose to eliminate. But then he is left with the dialectical contradiction that it was only by means of these mistakes that he arrived at the wisdom to be able to judge them to be such. As Hegel profoundly observed, the self-same maxims on the lips of a youth do not carry the same weight as when spoken by a man whose life's experience has filled them with meaning and content. They are the same and yet not the same. What was initially an abstract thought, with little or no real content, now becomes the product of mature reflection.

It was Hegel's genius to understand that the history of different philosophical schools was itself a dialectical process. He compares it to the life of a plant, going through different stages, that negate each other, but which, in their totality, represent the life of the plant itself:

> The more the ordinary mind takes the opposition between true and false to be fixed, the more is it accustomed to expect either agreement or contradiction with a given philosophical system, and only to see reason for the one or the other in any explanatory statement concerning such a system. It does not conceive the diversity of philosophical systems as the progressive evolution of truth; rather, it sees only contradiction in that variety. The bud disappears when the blossom breaks through, and we might say that the former is refuted by the latter; in the same way when the fruit comes, the blossom may be explained to be a false form of the plant's existence, for the fruit appears as its true nature in place of the blossom. These stages are not merely differentiated; they supplant one another as being incompatible with one another. But the ceaseless activity of their own inherent nature makes them at the same time moments of an organic unity, where they not merely do not contradict one another, but where one is as necessary as the other; and this equal necessity of all moments constitutes alone and thereby the life of the whole. [40]

The Dialectics of Capital

In the three volumes of *Capital*, Marx provides a brilliant example of how the dialectical method can be used to analyze the most fundamental processes in society. By so doing, he revolutionized the science of political economy, a fact which is not denied even by those economists whose views sharply conflict with those of Marx. So fundamental is the dialectical method to Marx's work, that Lenin went so far as to say that it was not possible to understand *Capital*, and especially its first chapter, without having read the whole of Hegel's *Logic!* This was undoubtedly an exaggeration. But what Lenin was driving at was the fact that Marx's *Capital* is itself a monumental object-lesson on how dialectics ought to be applied.

> If Marx did not leave behind him a 'Logic' (with a capital letter), he did leave the logic of *Capital*, and this ought to be utilized to the full in this question. In *Capital*, Marx applied to a single science logic, dialectics and the theory of knowledge of materialism [three words are not needed: it is one and the same thing] which has taken everything valuable in Hegel and developed it further. [41]

What method did Marx use in *Capital*? He did not impose the laws of dialectics upon economics but derived them from a long and painstaking study of all aspects of the economic process. He did not put forward an arbitrary schema and then proceed to make the facts fit into it but set out to uncover the laws of motion of capitalist production through a careful examination of the phenomenon itself. In his *Preface to the Critique of Political Economy*, Marx explains his method:

> I am omitting a general introduction which I had jotted down because on closer reflection any anticipation of results still to be proved appears to me to be objectionable, and the reader who on the whole desires to follow me must be resolved to ascend from the particular to the general. [42]

Capital represented a breakthrough, not only in the field of economics, but for social science in general. It has a direct relevance to the kind of discussions which are taking place among scientists at the present time. When Marx was alive, this discussion had already begun. At that time, scientists were obsessed with the idea of taking things apart and examining them in detail. This method is now referred to as "reductionism," although Marx and Engels, who were highly critical of it, called it the "metaphysical method." The mechanicists dominated physics for 150 years. Only now is the reaction against reductionism gathering steam. A new generation of scientists is setting itself the task of overcoming this heritage, and moving on to the formulation of new principles, in place of the old approximations.

It was thanks to Marx that the reductionist tendency in economics was routed in the middle of the last century. After *Capital*, such an approach was unthinkable. The "Robinson Crusoe" method of explaining political economy ("imagine two people on a desert island...") occasionally resurfaces in bad school text-books and vulgar attempts at popularization, but cannot be taken seriously. Economic crises and revolutions do not take place between two individuals on a desert island! Marx analyzes the capitalist economy, not as the sum-total of individual acts of exchange, but as a complex system, dominated by laws of its own which are as powerful as the laws of nature. In the same way, physicists are now discussing the idea of complexity, in the sense of a system in which the whole is not just a collection of elementary parts. Of course, it is useful to know, where possible, the laws which govern each individual part, but the complex system will be governed by new laws which are not merely extensions of the previous ones. This is precisely the method of Marx's *Capital* — the method of dialectical materialism.

Marx begins his work with an analysis of the basic cell of capitalist economy — the commodity. From this he explains how all the contradictions of capitalist society arise. Reductionism treats things like whole and part, particular and universal as mutually incompatible and exclusive, whereas they are completely inseparable, and interpenetrate and determine each other. In the first volume of *Capital*, Marx explains the twofold nature of commodities, as use-values and exchange-values. Most people see commodities exclusively as use-values, concrete, useful objects for the satisfaction of human wants. Use-values have always been produced in every type of human society.

However, capitalist society does strange things to use-values. It converts them into exchange-values — goods which are produced not directly for consumption, but for sale. Every commodity thus has two faces — the homely, familiar face of a use-value, and the mysterious, hidden face of an exchange-value. The former is directly linked to the physical properties of a particular commodity (we wear a shirt, drink coffee, drive a car, etc.). But exchange value cannot be seen, worn or eaten. It has no material being whatsoever. Yet it is the essential nature of a commodity under capitalism. The ultimate expression of exchange-value is money, the universal equivalent, through which all commodities express their value. These little pieces of green paper have no relation whatever to shirts, coffee or cars as such. They cannot be eaten, worn or driven. Yet such is the power they contain, and so universally is this recognized, that people will kill for them.

The dual nature of the commodity expresses the central contradiction of capitalist society — the conflict between wage-labor and capital. The worker thinks he sells his labor to the employer, but in fact what he sells is his labor power, which the capitalist uses as he sees fit. The surplus value thus extracted is the unpaid labor of the working class, the source of the accumulation of capital. It is this unpaid labor which maintains all the non-working members of society, through rent, interest, profits and taxation. The class struggle is really the struggle for the division of this surplus value.

Marx did not invent the idea of surplus value, which was known to previous economists like Adam Smith and Ricardo. But, by disclosing the central contradiction involved in it, he completely revolutionized political economy. This discovery can be compared to a similar process in the history of chemistry. Until the late 18th century, it was assumed that the essence of all combustion consisted in the separation from burning substances of a hypothetical thing called phlogiston. This theory served to explain most of the known chemical phenomena at the time. Then in 1774, the English scientist Joseph Priestley discovered something which he called "dephlogisticated air," which was later found to disappear whenever a substance was burned in it.

Priestley had, in fact, discovered oxygen. But he and other scientists were unable to grasp the revolutionary implications of this discovery. For a long time afterwards they continued to think in the old way. Later, the French chemist Lavoisier discovered that the new kind of air was really a chemical element, which did not disappear in the process of burning, but combined with the burnt substance. Although others had discovered oxygen, they did not know what they had discovered. This was the great discovery of Lavoisier. Marx played a similar role in political economy.

Marx's predecessors had discovered the existence of surplus value, but its real character remained shrouded in obscurity. By subjecting all previous theories, beginning with Ricardo, to a searching analysis, Marx discovered the real, contradictory nature of value. He examined all the relations of capitalist society, starting with the simplest form of commodity production and exchange, and following the process through all its manifold transformations, pursuing a strictly dialectical method.

Marx showed the relation between commodities and money, and was the first one to provide an exhaustive analysis of money. He showed how money is transformed into capital, demonstrating how this change is brought about through the buying and selling of labor power. This fundamental distinction between labor and labor power was the key that unlocked the mysteries of surplus value, a problem that Ricardo had been unable to solve. By establishing the difference between constant and variable capital, Marx was able to trace the entire process of the formation of capital in detail, and thus explain it, which none of his predecessors were able to do.

Marx's method throughout is rigorously dialectical, and follows quite closely the main lines traced by Hegel's *Logic*. This is explicitly stated in the Afterword to the Second German edition, where Marx pays a handsome tribute to Hegel:

> Whilst the writer pictures what he takes to be actually my method, in this striking and [as far as concerns my own application of it] generous way, what else is he picturing but the dialectic method?

> Of course the method of presentation must differ in form from that of inquiry. The latter has to appropriate the material in detail, to analyze its different forms of development, to trace out their inner connection. Only after this work is done, can the actual movement be adequately described. If this is done successfully, if the life of the subject-matter is ideally reflected as in a mirror, then it may appear as if we had before us a mere a priori construction...

> The mystifying side of Hegelian dialectic I criticized nearly thirty years ago, at a time when it was still the fashion. But just as I was working at the first volume of *Das Kapital*, it was the good pleasure of the peevish, arrogant, mediocre Epigonoi who now talk large in cultured Germany, to treat Hegel in the same way as the brave Moses Mendelssohn in Lessing's time treated Spinoza, i.e., a 'dead dog.' I

therefore openly avowed myself the pupil of that mighty thinker, and even here and there, in the chapter on the theory of value, coquetted with the modes of expression peculiar to him. The mystification which dialectic suffers in Hegel's hands, by no means prevents him from being the first to present its general form of working in a comprehensive and conscious manner. With him it is standing on its head. It must be turned right side up again, if you would discover the rational kernel within the mystical shell.

In its mystified form, dialectic become the fashion in Germany, because it seemed to transfigure and to glorify the existing state of things. In its rational form it is a scandal and abomination to bourgeoisdom and its doctrinaire professors, because it includes in its comprehension and affirmative recognition of the existing state of things, at the same time also, the recognition of the negation of that state, of its inevitable breaking up; because it regards every historically developed social form as in fluid movement, and therefore takes into account its transient nature not less than its momentary existence; because it lets nothing impose upon it, and is in its essence critical and revolutionary. [43]

4. FORMAL LOGIC AND DIALECTICS

The ability of men and women to think logically is the product of a lengthy process of social evolution. It antedates the invention of formal logic, not by thousands, but by millions of years. Locke already expressed this thought in the 17th century, when he wrote: *"God has not been so sparing to men as to make them barely two-legged creatures, and left it to Aristotle to make them rational." Behind Logic, according to Locke, stands "a naïve faculty to perceive the coherence or incoherence of its ideas."* [44]

The categories of logic did not drop from the clouds. These forms have taken shape in the course of the socio-historical development of humankind. They are elementary generalizations of reality, reflected in the minds of men and women. They are drawn from the fact that every object has certain qualities which distinguish it from other objects; that everything exists in certain relations to other things; that objects form larger classes, with which they share specific properties; that certain phenomena cause other phenomena, and so on.

To some extent, as Trotsky remarked, even animals possess the ability to reason and draw certain conclusions from a given situation. In higher mammals, and in particular the apes, this capacity is quite advanced, as the most recent research into bonobo chimpanzees strikingly reveal. However, while the capacity to reason may not be a monopoly of the human species, there is no doubt that, at least in our small corner of the universe, the ability to think rationally has reached its highest point so far in the development of the human intellect.

Abstraction is absolutely necessary. Without it, thought in general would be impossible. The question is: what sort of abstraction? When I abstract from reality, I concentrate on some aspects of a given phenomenon, and leave the others out of account. A good mapmaker, for instance, is not someone who reproduces every detail of every house and paving-stone, and every parked car. Such an amount of detail would destroy the very purpose of the map, which is to make available a convenient scheme of a town or other geographical area. Similarly, the brain early on learns to ignore certain sounds and concentrate on others. If we were not able to do this, the amount of information reaching our ears from all sides would overwhelm the mind completely. Language itself presupposes a high level of abstraction.

The ability to make correct abstractions, which adequately reflect the reality we wish to understand and describe, is the essential prerequisite for scientific thought. The abstractions of formal logic are adequate to express the real world only within quite narrow limits. But they are one-sided and static, and are hopelessly inadequate to deal with complex processes, particularly movement, change and contradictions. The concreteness of an object consists of the sum-total of its aspects and interrelationships, determined by its underlying laws. It is the task of science to uncover these laws, and to get as close as possible to this concrete reality. The whole purpose of cognition is to reflect the objective world and its underlying lawfulness and necessary relationships as faithfully as possible. As Hegel point out, *"the truth is always concrete."*

But here we have a contradiction. It is not possible to arrive at an understanding of the concrete world of nature without first resorting to abstraction. The word abstract comes from the Latin *"to take from."* By a process of abstraction, we take from the object under consideration certain aspects which we consider important, leaving others to one side. Abstract knowledge is necessarily one-sided because it expresses only one particular side of the phenomenon under consideration, isolated from that which determines the specific nature of the whole. Thus, mathematics deals exclusively with quanti-tative relations. Since quantity is an extremely important aspect of nature, the abstractions of mathematics have provided us with a powerful instrument for probing her secrets. For this reason, it is tempting to forget their real nature and limitations. Yet they remain one-sided, like all abstractions. We forget this at our peril.

Nature knows quality as well as quantity. To determine the precise relation between the two, and to show how, at a critical point, one turns into the other is absolutely necessary if we wish to understand one of the most fundamental processes in nature. This is one of the most basic concepts of dialectical as opposed to merely formal thought, and one of its most important contributions to science. The deep insights provided by this method, which was long decried as *"mysticism,"* are only now beginning to be understood and appreciated. One-sided abstract thought, as manifested in formal logic did a colossal disservice to science by excommunicating dialectics. But the actual results of science show that, in the last analysis, dialectical thinking is far closer to the real processes of nature than the linear abstractions of formal logic.

It is necessary to acquire a concrete understanding of the object as an integral system, not as isolated fragments; with all its necessary interconnections, not torn out of context, like a butterfly pinned to a collector's board; in its life and movement, not as something lifeless and static. Such an approach is in open conflict with the so-called "laws" of formal logic, the most absolute expression of dogmatic thought ever conceived, representing a kind of mental rigor mortis. But nature lives and breathes, and stubbornly resists the embraces of formalistic thinking. "A" is not equal to "A." Subatomic particles are and are not. Linear processes end in chaos. The whole is greater than the sum of its parts. Quantity changes into quality. Evolution itself is not a gradual process, but interrupted by sudden leaps and catastrophes. What can we do about it? Facts are stubborn things.

Without abstraction it is impossible to penetrate the object in "depth," to understand its essential nature and laws of motion. Through the mental work of abstraction, we are able to get beyond the immediate information provided by our senses (sense-perception), and probe deeper. We can break the object down into its constituent parts, isolate them, and study them in detail. We can arrive at an idealized, general conception of the object as a "pure" form, stripped of all secondary features. This is the work of abstraction, an absolutely necessary stage of the process of cognition.

> *Thought proceeding from the concrete to the abstract* [wrote Lenin], — *provided it is correct (and Kant, like all philosophers, speaks of correct thought) — does not get away from the truth but comes closer to it. The abstraction of matter, of a law of nature, the abstraction of value, etc., in short all scientific (correct, serious, not absurd) abstractions reflect nature more deeply, truly and completely. From living perception to abstract thought, and from this to practice, — such is the dialectical path of the cognition of truth, of the cognition of objective reality.* [45]

One of the main features of human thought is that it is not limited to what is, but also deals with what must be. We are constantly making all kinds of logical assumptions about the world we live in. This logic is not learned from books, but is the product of a long period of evolution. Detailed experiments have shown that the rudiments of this logic is acquired by a baby at a very young age, from experience. We reason that if something is true, then something else, for which we have no immediate evidence, must also be true. Such logical thought-processes take place millions of times all our waking hours, without us even being aware of them. They acquire the force of habit, and even the simplest actions in life would not be possible without them.

The elementary rules of thought are taken for granted by most people. They are a familiar part of life, and are reflected in many proverbs, such as *"you can't have your cake and eat it"* — a most important lesson for any child to learn! At a certain point, these rules were written down and systematized. This is the origin of formal logic, for which Aristotle must take the credit, along with so many other things. This was most valuable, since without a knowledge of the elementary rules of logic, thought runs the risk of becoming incoherent. It is necessary to distinguish black from white, and know the difference between a true statement and one that is false. The value of formal logic is, therefore, not in question. The problem is that the categories of formal logic, drawn from quite a limited range of experience and observation, are really valid only within these limits. They do, in fact, cover a great deal of everyday phenomena, but are quite inadequate to deal with more complex processes, involving movement, turbulence, contradiction, and the change from quality to quality.

In an interesting article entitled *The Origins of Inference*, which appeared in the anthology *Making Sense*, on the child's construction of the world, Margaret Donaldson draws attention to one of the problems of ordinary logic — its static character:

> *Verbal reasoning commonly appears to be about "states of affairs" — the world seen as static, in a cross-section of time. And considered in this way the universe appears to contain no incompatibility: things just are as they are. That object over there is a tree; that cup is blue; that man is taller than that man. Of course these states of affairs preclude infinitely many others, but how do we come to be aware of this? How does the idea of incompatibility arise in our minds? Certainly not directly from our impressions of things-as-they-are.*

The same book makes the valid point that the process of knowing is not passive, but active:

> *We do not sit around passively waiting for the world to impress its "reality" on us. Instead, as is now widely recognized, we get much of our most basic knowledge through taking action"* [46]

Human thought is essentially concrete. The mind does not readily assimilate abstract concepts. We feel most at home with what is immediately before our eyes, or at least with things that can be represented in a concrete way. It is as if the mind requires a crutch in the shape of images. On this, Margaret Donaldson remarks that *"even preschool children can frequently reason well about the events in the stories they hear. However, when we move beyond the bounds of human sense there is a dramatic difference. Thinking which does move beyond these bounds, so that it no longer operates within the supportive context of meaningful events, is often called 'formal' or 'abstract.'"* [47]

The initial process thus goes from the concrete to the abstract. The object is dismembered, analyzed, in order to obtain a detailed knowledge of its parts.

But there are dangers in this. The parts cannot be correctly understood apart from their relationship with the whole. It is necessary to return to the object as an integral system, and to grasp the underlying dynamics that condition it as a whole. In this way, the process of cognition moves from the abstract back to the concrete. This is the essence of the dialectical method, which combines analysis with synthesis, induction and deduction.

The whole swindle of idealism is derived from an incorrect understanding of the nature of abstraction. Lenin pointed out that the possibility of idealism is inherent in any abstraction. The abstract concept of a thing is counter-posed artificially to the thing itself. It is supposed not only to have an existence of its own, but is said to be superior to crude material reality. The concrete is portrayed as somehow defective, imperfect and impure, as opposed to the Idea which is perfect, absolute and pure. Thus reality is stood on its head.

The ability to think in abstractions marks a colossal conquest of the human intellect. Not only "pure" science, but also engineering would be impossible without abstract thought, which lifts us above the immediate, finite reality of the concrete example, and gives thought a universal character. The unthinking rejection of abstract thought and theory indicates the kind of narrow, Philistine mentality, which imagines itself to be "practical," but, in reality, is impotent. Ultimately, great advances in theory lead to great advances in practice. Nevertheless, all ideas are derived one way or another from the physical world, and, ultimately, must be applied back to it. The validity of any theory must be demonstrated, sooner or later, in practice.

In recent years there has been a healthy reaction against the mechanical reductionism, counter-posing the need for a holistic approach to science. The term holistic is unfortunate, because of its mystical associations. Nevertheless, in attempting to see things in their movement and interconnections, chaos theory undoubtedly comes close to dialectics. The real relationship between formal logic and dialectics is that between the type of thinking that takes things apart, and looks at them separately, and that which is also able to put them together again and make them work. If thought is to correspond to reality, it must be capable of grasping it as a living whole, with all its contradictions.

What is a Syllogism?

> *Logical thinking, formal thinking in general* [says Trotsky], *is constructed on the basis of the deductive method, proceeding from a more general syllogism through a number of premises to the necessary conclusion. Such a chain of syllogisms is called a sorites.* [48]

Aristotle was the first one to write a systematic account of both dialectics and formal logic, as methods of reasoning. The purpose of formal logic was to provide a framework to distinguish valid from invalid arguments. This he did in the form of syllogisms. There are different forms of syllogism, which are really variations on the same theme.

Aristotle in his *Organon*, names ten categories — substance, quantity, quality, relation, place, time, position, state, action, passion, which form the basis of the dialectical logic, later given its full expression in the writings of Hegel. This side of Aristotle's work on logic is frequently ignored. Bertrand Russell, for example, considered these categories to be meaningless. But since logical positivists like Russell have written off practically the whole history of philosophy (except the bits and pieces that coincide with their dogmas) as "meaningless," this should neither surprise nor trouble us too much.

The syllogism is a method of logical reasoning, which may be variously described. The definition given by Aristotle himself was as follows: *"A discourse in which, certain things being stated, something other than what is stated follows of necessity from their being so." The simplest definition is given by A. A. Luce: "A syllogism is a triad of connected propositions, so related that one of them, called the Conclusion, necessarily follows from the other two, which are called the Premises."* [49]

The mediaeval Schoolmen focused their attention on this kind of formal logic which Aristotle developed in The Prior and Posterior Analytics. It is in this form that Aristotle's logic came down from the Middle Ages. In practice, the syllogism consists of two premises and a conclusion. The subject and the predicate of the conclusion each occur in one of the premises, together with a third term (the middle) that is found in both premises, but not in the conclusion. The predicate of the conclusion is the major term; the premise in which it is contained is the major premise; the subject of the conclusion is the minor term; and the premise in which it is contained is the minor premise. For example,

a) All men are mortal (Major premise).

b) Caesar is a man (Minor premise).

c) Therefore, Caesar is mortal (Conclusion).

This is called an affirmative categorical statement. It gives the impression of being a logical chain of argument, in which each stage is derived inexorably from the previous one. But actually, this is not the case, because "Caesar" is already included in *"all men."* Kant, like Hegel, regarded the syllogism (that *"tedious doctrine,"* as he called it) with contempt. For him, it was *"nothing more than an artifice"* in which the conclusions were already surreptitiously introduced into the premises to give a false appearance of reasoning. [50]

Another type of syllogism is conditional in form (if...then), for example: *"If an animal is a tiger, it is a carnivore."* This is just another way of saying the same thing as the affirmative categorical statement, i.e., all tigers are carnivores. The same in relation to the negative form — *"If it's a fish, it's not a mammal"* is just another way of saying *"No fishes are mammals."* The formal difference conceals the fact that we have not really advanced a single step.

What this really reveals is the inner connections between things, not just in thought, but in the real world. "A" and "B" are related in certain ways to "C" (the middle) and the premise, therefore, they are related to each other in the conclusion. With great profundity and insight, Hegel showed that what the syllogism showed was the relation of the particular to the universal. In other words, the syllogism itself is an example of the unity of opposites, the contradiction par excellence, and that, in reality, all things are a *"syllogism."*

The heyday of the syllogism was in the Middle Ages, when the Schoolmen devoted their entire lives to endless disputations on all manner of obscure theological questions, like the sex of angels. The labyrinthine constructions of formal logic made it appear that they were really involved in a profound discussion, when, in fact, they were arguing about nothing at all. The reason for this lies in the nature of formal logic itself. As the name suggests, it is all about form. The question of the content does not enter into it. This is precisely the chief defect of formal logic, and its Achilles' heel.

By the time of the Renaissance, that great re-awakening of the human spirit, dissatisfaction with Aristotelian logic was widespread. There was a growing reaction against Aristotle, which was not really fair to this great thinker, but stemmed from the fact that the Church had suppressed all that was worthwhile in his philosophy, and preserved only a lifeless caricature. For Aristotle, the syllogism was only part of the process of reasoning, and not necessarily the most important part. Aristotle also wrote on the dialectic, but this aspect was forgotten. Logic was deprived of all life, and turned, in Hegel's phrase, into *"the lifeless bones of a skeleton."*

The revulsion against this lifeless formalism was reflected in the movement towards empiricism, which gave a tremendous impetus to scientific investigation and experiment. However, it is not possible to dispense altogether with forms of thought, and empiricism from the beginning carried the seeds of its own destruction. The only viable alternative to inadequate and incorrect methods of reasoning is to develop adequate and correct ones.

By the end of the Middle Ages the syllogism was discredited everywhere, and subjected to ridicule and abuse. Rabelais, Petrarch and Montaigne all denounced it. But it continued to trundle along, especially in those Catholic lands, untouched by the fresh winds of the Reformation. By the end of the 18th century, logic was in such a bad state that Kant felt obliged to launch a general criticism of the old thought forms in his *Critique of Pure Reason.*

Hegel was the first one to subject the laws of formal logic to a thoroughgoing critical analysis. In this, he was completing the work commenced by Kant. But whereas Kant only showed the inherent deficiencies and contradictions of traditional logic, Hegel went much further, working out a completely different approach to logic, a dynamic approach, which would include movement and contradiction, which formal logic is powerless to deal with.

Does Logic Teach How to Think?

Dialectics does not pretend to teach people to think. That is the pretentious claim of formal logic, to which Hegel ironically replied that logic no more teaches you to think than physiology teaches you to digest! Men and women thought, and even thought logically, long before they ever heard of logic. The categories of logic, and also dialectics, are derived from actual experience. For all their pretensions, the categories of formal logic do not stand above the crude world of material reality, but are only empty abstractions taken from reality comprehended in a one-sided and static manner, and then arbitrarily applied back to it.

By contrast, the first law of the dialectical method is absolute objectivity. In every case, it is necessary to discover the laws of motion of a given phenomenon by studying it from every point of view. The dialectical method is of great value in approaching things correctly, avoiding elementary philosophical blunders, and making sound scientific hypotheses. In view of the astonishing amount of mysticism that has emerged from arbitrary hypotheses, above all in theoretical physics, this is no mean advantage! But the dialectical method always seeks to derive its categories from a careful study of the facts and processes, not to force the facts into a rigid preconceived straitjacket:

> *We all agree* [wrote Engels], *that in every field of science, in natural as in historical science, one must proceed from the given facts, in natural science therefore from the various material forms and the various forms of motion of matter; that therefore in theoretical natural science too the interconnections are not to be built into the facts but to be discovered in them, and when discovered to be verified as far as possible by experiment.* [51]

Science is founded on the search for general laws which can explain the workings of nature. Taking its starting point as experience, it does not confine itself to the mere collection of facts, but seeks to generalize on the basis of experience, going from the particular to the universal. The history of science is characterized by an ever-deepening process of approximation. We get closer and closer to the truth, without ever knowing the *"whole truth."* Ultimately, the test of scientific truth is experiment. *"Experiment,"* says Feynman, *"is the sole judge of scientific 'truth.'"* [52]

The validity of forms of thought must, in the last analysis, depend on whether they correspond to the reality of the physical world. This cannot be established a priori, but must be demonstrated through observation and experiment. Formal logic, in contrast to all the natural sciences, is not empirical. Science derives its data from observation of the real world. Logic is supposed to be a priori, unlike all the subject matter with which it deals. There is a glaring contradiction here between form and content. Logic is not supposed to be derived from the real world, yet it is constantly applied to the facts of the real world. What is the relationship between the two sides?

Kant long ago explained that the forms of logic must reflect objective reality, or they would be entirely meaningless:

When we have reason to consider a judgment necessarily universal...we must consider it objective also, that is, that it expresses not merely a reference of our perception to a subject, but a quality of the object. For there would be no reason for the judgments of other men necessarily agreeing with mine, if it were not the unity of the object to which they all refer, and with which they accord; hence they must all agree with one another. [53]

This idea was developed further by Hegel, who removed the ambiguities present in Kant's theory of knowledge and logic, and finally put on a sound basis by Marx and Engels:

Logical schemata [Engels explains], can only relate to forms of thought; but what we are dealing with here are only forms of being, of the external world, and these forms can never be created and derived by thought out of itself, but only from the external world. But with this the whole relationship is inverted: the principles are not the starting point of the investigation, but its final result; they are not applied to nature and human history, but abstracted from them; it is not nature and the realm of humanity which conform to these principles, but the principles are only valid in so far as they are in conformity with nature and history. [54]

Limits of the Law of Identity

It is an astonishing fact that the basic laws of formal logic worked out by Aristotle have remained fundamentally unchanged for over two thousand years. In this period, we have witnessed a continuous process of change in all spheres of science, technology and human thought. And yet scientists have been content to continue to use essentially the same methodological tools that were used by the mediaeval School men in the days when science was still on the level of alchemy.

Given the central role played by formal logic in Western thought, it is surprising how little attention is paid to its real content, meaning and history. It is normally taken as something given, self-evident, and fixed for all time. Or it is presented as a convenient convention which reasonable people agree upon, in order to facilitate thought and discourse, rather as people in polite social circles agree upon good table manners. The idea is put forward that the laws of logic are entirely artificial constructions, made up by logicians, in the belief that they will have some application in some field of thought, where they will reveal some truth or other. But why should the laws of logic have any bearing upon anything, if they are only abstract constructions, the arbitrary imaginings of the brain?

On this idea, Trotsky commented ironically:

"To say that people have come to an agreement about the syllogism is almost like saying, or more correctly it is exactly the same as saying, that people came to an agreement to have nostrils in their noses. The syllogism is no less an objective product of organic development, i.e., the biological, anthropological, and social development of humanity than are our various organs, among them our organ of smell." In reality, formal logic is ultimately derived from experience, just as any other way of thinking. From their experience, humans draw certain conclusions, which they apply in their daily life. This applies even to animals, though at a different level.

"The chicken knows that grain is in general useful, necessary, and tasty. It recognizes a given piece of grain as that grain — of the wheat — with which it is acquainted and hence draws a logical conclusion by means of its beak. The syllogism of Aristotle is only an articulated expression of those elementary mental conclusions which we observe at every step among animals." [55]

Trotsky once said that the relationship between formal logic and dialectics was similar to the relationship between lower and higher mathematics. The one does not deny the other and continues to be valid within certain limits. Likewise, Newton's laws, which were dominant for a hundred years, were shown to be false in the world of subatomic particles. More correctly, the old mechanistic physics, which was criticized by Engels, was shown to be one-sided and of limited application.

"The dialectic," writes Trotsky, *"is neither fiction nor mysticism, but a science of the forms of our thinking insofar as it is not limited to the daily problems of life but attempts to arrive at an understanding of more complicated and drawn-out processes."* [56]

The most common method of formal logic is that of deduction, which attempts to establish the truth of its conclusions by meeting two distinct conditions a) the conclusion must really flow from the premises; and b) the premises themselves must be true. If both conditions are met, the argument is said to be valid. This is all very comforting. We are here in the familiar and reassuring realm of common sense. *"True or false?"* *"Yes or no?"* Our feet are firmly on the ground. We appear to be in possession of *"the truth, the whole truth, and nothing but the truth."* There is not a lot more to be said. Or is there?

Strictly speaking, from the standpoint of formal logic, it is a matter of indifference whether the premises are true or false. As long as the conclusions can be correctly drawn from its premises, the inference is said to be deductively valid. The important thing is to distinguish between valid and invalid inferences. Thus, from the standpoint of formal logic, the following assertion is deductively valid: All scientists have two heads. Einstein was a scientist. Therefore, Einstein had two heads. The validity of the inference does not depend upon the subject matter in the slightest. In this way, the form is elevated above the content.

In practice, of course, any mode of reasoning that did not demonstrate the truth of its premises would be worse than useless. The premises must be shown to be true. But this leads us into a contradiction. The process of validating one set of premises automatically raises a new set of questions, which in turn need to be validated. As Hegel points out, every premise gives rise to a new syllogism, and so on ad infinitum. So that what appeared to be very simple turns out to be extremely complex, and contradictory.

The biggest contradiction of all lies in the fundamental premises of formal logic itself. While demanding that everything else under the sun must justify itself in the High Court of the Syllogism, logic becomes utterly confused when asked to justify its own presuppositions. It suddenly loses all its critical faculties, and resorts to appeals to belief, common sense, the *"obvious,"* or, the final philosophical get-out clause — a priori. The fact is that the so-called axioms of logic are unproved formulas. These are taken as the starting point, from which all further formulae (theorems) are deduced, exactly as in classical geometry, where the starting point is provided by Euclid's principles. They are assumed to be correct, without any proof whatsoever, i.e., we just have to take them on trust.

But what if the basic axioms of formal logic turn out to be false? Then we would be in just the same position as when we gave poor Mr. Einstein an additional head. Is it conceivable that the eternal laws of logic might be flawed? Let us examine the matter more closely. The basic laws of formal logic are:

1) The law of identity ("A" = "A").

2) The law of contradiction ("A" does not equal "not-A").

3) The law of the excluded middle ("A" does not equal "B").

These laws, at first sight, seem eminently sensible. How could anyone quarrel with them? Yet closer analysis shows that these laws are full of problems and contradictions of a philosophical nature. In his *Science of Logic*, Hegel provides an exhaustive analysis of the Law of Identity, showing it to be one-sided and, therefore, incorrect.

Firstly, let us note that the appearance of a necessary chain of reasoning, in which one step follows from another, is entirely illusory. The law of contradiction merely restates the law of identity in a negative form. The same is true of the law of the excluded middle. All we have is a repetition of the first line in different ways. The whole thing stands or falls on the basis of the law of identity ("A" = "A"). At first sight this is incontrovertible, and, indeed, the source of all rational thought. It is the Holy of Holies of Logic, and not to be called into question. Yet called into question it was, and by one of the greatest minds of all time.

There is a story by Hans-Christian Andersen called *The Emperor's New Clothes*, in which a rather foolish emperor is sold a new suit by a swindler, which is supposed to be very beautiful, but invisible. The gullible emperor goes about in his fine new suit, which everyone agrees is exquisite, until one day a little boy points out that the emperor is, in fact, stark naked. Hegel performed a comparable service to philosophy in his critique of formal logic. Its defenders have never forgiven him for it.

The so-called law of identity is, in fact, a tautology. Paradoxically, in traditional logic, this was always regarded as one of the most glaring mistakes which can be committed in defining a concept. It is a logically untenable definition which merely repeats in other words what is already contained in the part to be defined. Let us put this more concretely. A teacher asks his pupil what a cat is, and the pupil proudly informs him that a cat is — a cat. Such an answer would not be considered very intelligent. After all, a sentence is generally intended to say something, and this sentence tells us nothing at all. Yet this not very bright scholar's definition of a feline quadruped is a perfect expression of the law of identity in all its glory. The young person concerned would immediately be sent to the bottom of the class. Yet for over two thousand years, the most learned professors have been content to treat it as the most profound philosophical truth.

All that the law of identity tells us about something is that it is. We do not get a single step further. We remain on the level of the most general and empty abstraction. For we learn nothing about the concrete reality of the object under consideration, its properties and relationships. A cat is a cat; I am myself; you are you; human nature is human nature; things are as they are. The emptiness of such assertions stands out in all its uncouthness. It is the consummate expression of one-sided, formalistic, dogmatic thinking.

Is the law of identity invalid, then? Not entirely. It has its applications, but these are far more limited in scope than what one might think. The laws of formal logic can be useful in clarifying certain concepts, analyzing, labeling, cataloguing, defining. It has the merit of neatness and tidiness. This has its place. For normal, simple, everyday phenomena, it holds good. But when dealing with more complex phenomena, involving movement, sudden leaps, qualitative changes, it becomes wholly inadequate, and, in fact, breaks down completely.

The following extract by Trotsky brilliantly sums up Hegel's line of argument in relation to the law of identity:

I will here attempt to sketch the substance of the problem in a very concise form. The Aristotelian logic of the simple syllogism starts from the proposition that "A" is equal to "A." This postulate is accepted as an axiom for a multitude of practical human actions and elementary generalizations. But in reality "A" is not equal to "A." This is easy to prove if we observe these two letters under a lens — they are quite different from each other. But, one can object, the question is not of the size or the form of the letters, since they are only symbols for equal quantities, for instance, a pound of sugar. The objection is beside the point; in reality a pound of sugar is never equal to a pound of sugar — a more delicate scale always discloses a difference. Again one can object: but a pound of sugar is equal to itself. Neither is this true — all bodies change uninterruptedly in size, weight, color, etc. They are never equal to themselves. A sophist will respond that a pound of sugar is equal to itself "at any given moment." Aside from the extremely dubious practical value of this "axiom," it does not withstand theoretical criticism either. How should we really conceive the word "moment"? If it is an infinitesimal interval of time, then a pound of sugar is subjected during the course of that "moment" to inevitable changes. Or is the "moment" a purely mathematical abstractions, that is, a zero of time? But everything exists in time; and existence itself is an uninterrupted process of transformation; time is consequently a fundamental element of existence. Thus the axiom "A" is equal to "A" signifies that a thing is equal to itself if it does not change, that is, if is does not exist.

At first glance it could seem that these "subtleties" are useless. In reality they are of decisive significance. The axiom "A" is equal to "A" appears on one hand to be the point of departure for all the errors in our knowledge. To make use of the axiom "A" is equal to "A" with impunity is possible only within certain limits. When quantitative changes in "A" are negligible for the task at hand then we can presume that "A" is equal to "A." This is, for example, the manner in which a buyer and a seller consider a pound of sugar. We consider the temperature of the sun likewise. Until recently we considered the buying power of the dollar in the same way. But quantitative changes beyond certain limits become converted into qualitative. A pound of sugar subjected to the action of water or kerosene ceases to be a pound of sugar. A dollar in the embrace of a president ceases to be a dollar. To determine at the right moment the critical point where quantity changes into quality is one of the most important and difficult tasks in all the spheres of knowledge including sociology...

Dialectical thinking is related to vulgar thinking in the same way that a motion picture is related to a still photograph. The motion picture does not outlaw the still photograph but combines a series of them according to the laws of motion. Dialectics does not deny the syllogism, but teaches us to combine syllogisms in such a way as to bring our understanding closer to the eternally changing reality. Hegel in his Logic established a series of laws: change of quantity into quality, development through contradictions, conflict of content and form, interruption of continuity, change of possibility into inevitability, etc., which are just as important for theoretical thought as is the simple syllogism for more elementary tasks. [57]

Similarly with the law of the excluded middle, which asserts that it is necessary either to assert or deny, that a thing must be either black or white, either alive or dead, either "A" or "B". It cannot be both at the same time. For normal everyday purposes, we can take this to be true. Indeed, without such assumptions, clear and consistent thought would be impossible. Moreover, what appear to be insignificant errors in theory sooner or later make themselves felt in practice, often with disastrous results. In the same way, a hairline crack in the wing of a jumbo jet may seem insignificant, and, indeed, at low speeds may pass unnoticed. At very high speeds, however, this tiny error can provoke a catastrophe. In *Anti-Dühring*, Engels explains the deficiencies of the so-called law of the excluded middle:

> *To the metaphysician* [wrote Engels], things and their mental images, ideas, are isolated, to be considered one after the other and apart from each other, fixed, rigid objects of investigation given once for all. He thinks in absolutely unmediated antitheses. "His communication is 'yea, yea; nay, nay'; for whatsoever is more than these cometh of evil." For him a thing either exists or does not exist; a thing cannot at the same time be itself and something else. Positive and negative absolutely exclude one another; cause and effect stand in a rigid antithesis one to the other.
>
> At first sight this way of thinking seems to us most plausible because it is that of so-called sound common sense. Yet sound common sense, respectable fellow that he is in the homely realm of his own four walls, has very wonderful adventures directly he ventures out into the wide world of research. The metaphysical mode of thought, justifiable and even necessary as it is in a number of domains whose extent varies according to the nature of the object, invariably bumps into a limit sooner or later, beyond which it becomes one-sided, restricted, abstract, lost in insoluble contradictions, because in the presence of individual things it forgets their connections; because in the presence of their existence it forgets their coming into being and passing away; because in their state of rest if forgets their motion. It cannot see the wood for the trees. For everyday purposes we know and can definitely say, e.g., whether an animal is alive or not. But, upon closer inquiry, we find that this is sometimes a very complex question, as the jurists very well know. They have cudgeled their brains in vain to discover a rational limit beyond which the killing of the child in its mother's womb is murder. It is just as impossible to determine the moment of death, for physiology proves that death is not a sudden instantaneous phenomenon, but a very protracted process.
>
> *In like manner, every organic being is every moment the same and not the same; every moment it assimilates matter supplied from without and gets rid of other matter; every moment some cells of its body die and others build themselves anew; in a longer or shorter time the matter of its body is completely renewed and is replaced by other molecules of matter, so that every organic being is always itself, and yet something other than itself.* [58]

The relationship between dialectics and formal logic can be compared to the relationship between quantum mechanics and classical mechanics. They do not contradict but complement each other. The laws of classical mechanics still hold good for an immense number of operations. However, they cannot be adequately applied to the world of subatomic particles, involving infinitesimally small quantities and tremendous velocities. Similarly, Einstein did not replace Newton, but merely exposed the limits beyond which Newton's system did not work.

Formal logic (which has acquired the force of popular prejudice in the form of "*common sense*") equally holds good for a whole series of everyday experiences.

However, the laws of formal logic, which set out from an essentially static view of things, inevitably break down when dealing with more complex, changing and contradictory phenomena. To use the language of chaos theory, the "linear" equations of formal logic cannot cope with the turbulent processes which can be observed throughout nature, society and history. Only the dialectical method will suffice for this purpose.

Logic and the Subatomic World

The deficiencies of traditional logic have been grasped by other philosophers, who are very far from the dialectical standpoint. In general, in the Anglo-Saxon world, there has traditionally been a greater inclination towards empiricism, and inductive reasoning. Nevertheless, science still requires a philosophical framework which will enable it to assess its results and guide its steps through the confused mass of facts and statistics, like Ariadne's thread in the labyrinth. Mere appeals to *"common sense,"* or the *"facts,"* will not suffice.

Syllogistic thinking, the abstract deductive method, is very much in the French tradition, especially since Descartes. The English tradition was altogether different, being heavily influenced by empiricism. From Britain, this school of thought was early on imported to the United States, where it sunk deep roots. Thus, the formal-deductive mode of thought was not at all characteristic of the Anglo-Saxon intellectual tradition. *"On the contrary,"* wrote Trotsky, *"it is possible to say that this [school of] thought is distinguished by a sovereign-empirical contempt for the pure syllogism, which did not prevent the English from making colossal conquests in many spheres of scientific investigation. If one really thinks this through as one should, then it is impossible not to arrive at the conclusion that the empirical disregard for the syllogism is a primitive form of dialectical thinking."*

Empiricism historically played both a progressive role (in the struggle against religion and mediaeval dogmatism) and a negative one (an excessively narrow interpretation of materialism, resistance to broad theoretical generalizations). Locke's famous assertion that there is nothing in the intellect which is not derived from the senses contains the germ of a profoundly correct idea, but presented in a one-sided way, which could, and did, have the most harmful consequences on the future development of philosophy. In relation to this, Trotsky wrote shortly before his assassination:

> *"We do not know anything about the world except what is provided through experience." This is correct if one does not understand experience in the sense of the direct testimony of our individual five senses. If we reduce the matter to experience in the narrow empirical sense, then it is impossible for us to arrive at any judgment concerning either the origin of the species or, still less, the formation of the earth's crust. To say that the basis for everything is experience is to say too much or to say nothing at all. Experience is the active interrelationship between subject and object. To analyze*

experience outside this category, i.e., outside the objective material milieu of the investigator who is counter-posed to it and who from another standpoint is a part of this milieu — to do this is to dissolve experience in a formless unity where there is neither object nor subject but only the mystical formula of experience. "Experiment" or "experience" of this kind is peculiar only to a baby in its mother's womb, but unfortunately the baby is deprived of the opportunity to share the scientific conclusions of its experiment. [59]

The uncertainty principle of quantum mechanics cannot be applied to ordinary objects, but only to atoms and subatomic particles. Subatomic particles obey different laws to those of the "ordinary" world. They move at incredible speeds, 1,500 meters per second, for example. They can move in different directions at the same time. Given this situation, the forms of thought which apply to everyday experience are no longer valid. Formal logic is useless. Its black and white, yes-or-no, take it or leave it categories have no point of contact with this fluid, unstable and contradictory reality. All we can do is to say that it is probably such and such a motion, with an infinite number of possibilities. Far from proceeding from the premises of formal logic, quantum mechanics violates the Law of Identity by asserting the *"non-individuality"* of individual particles. The Law of Identity cannot apply at this level, because the "identity" of individual particles cannot be fixed. Hence the lengthy controversy over *"wave"* or *"particle."* It could not be both! Here "A" turns out to be "not-A," and "A" can indeed be also "B." Hence, the impossibility of "fixing" an electron's position and velocity in the neat and absolute manner of formal logic. That is a serious problem for formal logic and "common sense," but not for dialectics or for quantum mechanics. An electron has both the qualities of a wave and a particle, and this has been experimentally demonstrated.

In 1932, Heisenberg suggested that the protons inside the nucleus were held together by something he called exchange force. This implied that protons and neutrons were constantly exchanging identity. Any given particle is in a constant state of flux, changing from a proton into a neutron and back again. Only in this way is the nucleus held together. Before a proton can be repelled by another proton, it changes into a neutron, and vice versa. This process in which particles are changed into their opposites takes place uninterruptedly, so that it is impossible to say at any given moment whether a particle is a proton or a neutron. In fact it is both — it is and is not.

The exchange of identities between electrons does not mean a simple change of position, but a complicated process where electron "A" interpenetrates with electron "B" to produce a *"mix"* of, say, 60% "A" and 40% "B" and vice versa. Later, they may have completely exchanged identities, with all *"A"* here and all *"B"* there. The flow would then be reversed in a permanent oscillation, involving a rhythmic interchange of the electrons' identities, which goes on indefinitely.

The old rigid, fixed Law of Identity vanishes altogether in this kind of pulsating identity-in-difference, which underlies all being, and received its scientific expression in Pauli's principle of exclusion.

Thus, two and a half millennia later, Heraclitus' principle "everything flows" turns out to be true - literally. Here we have, not only a state of unceasing change and motion, but also a process of universal interconnection, and the unity and interpenetration of opposites. Not only do electrons condition each other, but they actually pass into each other and become transformed into each other. How far removed from the static, unchanging idealist universe of Plato! How does one fix the position of an electron? By looking at it. And how to determine its momentum? By looking at it twice. But in that time, even in an infinitesimally small space of time, the electron has changed, and is no longer what it was. It is something else. It is both a particle (a *"thing,"* a *"point"*) and a wave (a *"process,"* movement, becoming). It is and is not. The old black and white method of formal logic used in classical mechanics cannot give results here because of the very nature of the phenomenon.

In 1963, it was suggested by Japanese physicists that the extremely small particle known as the neutrino changed its identity as it traveled through space at very high speeds. At one point it was an electron-neutrino, at another, a muon-neutrino, at another, a tauon-neutrino, and so on. If this is true, the law of identity, which has already been thoroughly battered, can be said to have received the final coup de grace. Such a rigid, black-and-white conception is clearly out of its depth when confronted with any one of the complex and contradictory phenomena of nature described by modern science.

Modern Logic

In the 19th century, there were a number of attempts to bring logic up to date (George Boyle, Ernst Schröder, Gotlob Frege, Bertrand Russell and A. N. Whitehead). But, apart from the introduction of symbols, and a certain tidying up, there is no real change here. Great claims are made, for example by the linguistic philosophers, but there are not many grounds for them. Semantics (which deals with the validity of an argument) was separated from syntax (which deals with the deductibility of the conclusions from axioms and premises). This is supposed to be something new, when, in reality, it is merely a re-hash of the old division, well known to the ancient Greeks, between logic and rhetoric. Modern logic is based on the logical relations among whole sentences. The center of attention has moved away from the syllogism towards hypothetical and disjunctive arguments. This is hardly a breathtaking leap. One can begin with sentences (judgments) instead of syllogisms. Hegel did this in his

Logic. Rather than a great revolution in thought, it is like re-shuffling cards in a pack.

Using a superficial and inexact analogy with physics, the so-called "*atomic method*" developed by Russell and Wittgenstein (and later repudiated by the latter) tried to divide language into its "atoms." The basic atom of language is supposed to be the simple sentence, out of which compound sentences are constructed. Wittgenstein dreamed of developing a "formal language" for every science — physics, biology, even psychology. Sentences are subjected to a "*truth test*" based on the old laws of identity, contradiction and the excluded middle. In reality, the basic method remains exactly the same. The "*truth value*" is a question of "*either...or,*" "*yes or no,*" "*true or false.*" The new logic is referred to as the propositional calculus. But the fact is that this system cannot even deal with arguments formerly handled by the most basic (categorical) syllogism. The mountain has labored, and brought forth a mouse.

The fact is that not even the simple sentence is really understood, although it is supposed to be the linguistic equivalent of the "*building-blocks of matter.*" Even the simplest judgment, as Hegel points out, contains a contradiction. "*Caesar is a man,*" "*Fido is a dog,*" "*the tree is green,*" all state that the particular is the universal. Such sentences seem simple, but in fact are not. This is a closed book for formal logic, which remains determined to banish all contradictions, not only from nature and society, but from thought and language as well. Propositional calculus sets out from exactly the same basic postulates as those worked out by Aristotle in the 4th century B.C., namely, the law of identity, the law of (non-)contradiction, the law of the excluded middle, to which is added the law of double negation. Instead of being written with normal letters, they are expressed in symbols, thus:

a) $p = p$

b) $p = \sim p$

c) $p \lor = \sim p$

d) $\sim (p \sim p)$

All this looks very nice, but makes not the slightest difference to the content of the syllogism. Moreover, symbolic logic itself is not a new idea. In the 1680s, the ever-fertile brain of the German philosopher Leibniz created a symbolic logic, although he never published it.

The introduction of symbols into logic does not carry us a single step further, for the very simple reason that they, in turn, must sooner or later be translated into words and concepts. They have the advantage of being a kind of shorthand, more convenient for some technical operations, computers and so on, but the content remains exactly as before. The bewildering array of mathematical symbols is accompanied by a truly Byzantine jargon, which seems deliberately designed to make logic inaccessible to ordinary mortals, just as the priest-castes of Egypt and Babylon used secret words and occult symbols to keep their knowledge to themselves. The only difference is that they actually did know things that were worth knowing, like the movements of the heavenly bodies, something which cannot be said of modern logicians.

Terms such as *"monadic predicates,"* *"qualifiers,"* *"individual* variables," and so on and so forth, are designed to give the impression that formal logic is a science to be reckoned with, since it is quite unintelligible to most people. Sad to say, the scientific value of a body of beliefs is not directly proportionate to the obscurity of its language. If that were the case, every religious mystic in history would be as great a scientist as Newton, Darwin and Einstein, all rolled into one.

In Moliere's comedy, *Le Bourgeois Gentilhomme*, M. Jourdain was surprised to be told that he had been talking prose all his life, without realizing it. Modern logic merely repeats all the old categories, but throws in a few symbols and fancy-sounding terms, in order to hide the fact that absolutely nothing new is being said. Aristotle used *"monadic predicates"* (expressions that attribute a property to an individual) a long time ago. No doubt, like M. Jourdain, he would have been delighted to discover that he had been using Monadic Predicates all the time, without knowing it. But it would not have made a scrap of difference to what he was actually doing. The use of new labels does not alter the contents of a jar of jam. Nor does the use of jargon enhance the validity of outworn forms of thought.

The sad truth is that, in the 20th century formal logic has reached its limits. Every new advance of science deals it yet another blow. Despite all the formal changes, the basic laws remain the same. One thing is clear. The developments of formal logic over the past hundred years, first by propositional calculus (p.c.), then by lower predicate calculus (l.p.c.) has carried the subject to such a point of refinement that no further development is possible. We have reached the most comprehensive system of formal logic, so that any other additions will certainly not add anything new. Formal logic has said all that it has to say. If the truth were to be told, it reached this stage quite some time ago.

Recently, the ground has shifted from argument to deducing conclusions. How are the *"theorems of logic derived"*? This is quite shaky ground. The basis of formal logic has always been taken for granted in the past. A thorough investigation of the theoretical grounds of formal logic would inevitably result in transforming them into their opposite. Arend Heyting, the founder of the Intuitionist School of mathematics, denies the validity of some of the proofs used in classical mathematics. However, most logicians cling desperately to the old laws of formal logic, like a drowning man clutching at a straw:

"We do not believe that there is any non-Aristotelian logic in the sense in which there is a non-Euclidean geometry, that is, a system of logic in which the contraries of the Aristotelian principles of contradiction and the excluded middle are assumed to be true, and valid inferences drawn from them." [60]

There are two main branches of formal logic today — propositional calculus and predicate calculus. They all proceed from axioms, which are assumed to be true *"in all possible worlds,"* under all circumstances. The fundamental test remains freedom from contradiction. Anything contradictory is deemed to be *"not valid."* This has a certain application, for example, in computers, which are geared to a simple yes or no procedure. In reality, however, all such axioms are tautologies. These empty forms can be filled with almost any content. They are applied in a mechanical and external fashion to any subject. When it comes to simple linear processes, they do their work tolerably well. This is important, because a great many of the processes in nature and society do, in fact, work in this way. But when we come to more complex, contradictory, non-linear phenomena, the laws of formal logic break down. It immediately becomes evident that, far from being universal truths valid "in all possible worlds," they are, as Engels explained, quite limited in their application, and quickly find themselves out of their depth in a whole range of circumstances. Moreover, these are precisely the kind of circumstances which have occupied the attention of science, especially the most innovative parts of it, for most of the 20th century.

NOTES PART ONE

For reasons of convenience, where the same work is cited several times in immediate sequence we have placed the reference number at the end of the last quote.

1 Karl Marx and Frederick Engels, *Selected Correspondence, Letter to Bloch, 21st-22nd September 1890*, henceforth referred to as *MESC*.
2 *The Economist*, 9th January 1982.
3 W. Rees-Mogg, *The Great Reckoning, How the World Will Change in the Depression of the 1990s*, p. 445.
4 Ibid., p. 27, our emphasis.
5 *The Guardian*, 9th March, 1995.
6 Gordon Childe, *What Happened in History*, p. 19.
7 Ibid., pp. 19-20.
8 Sir James Frazer, *The Golden Bough*, p. 10.
9 Ibid., p. 105.
10 Ludwig Feuerbach, *The Essence of Christianity*, p. 5.
11 Aristotle, *Metaphysics*, p. 53.
12 I. Prigogine and I. Stengers, *Order Out of Chaos, Man's New Dialogue with Nature*, p. 4.
13 Quoted in Margaret Donaldson, *Children's Minds*, p. 84.
14 *Oeconomicus*, iv, 203, quoted in B. Farrington, *Greek Science*, pp. 28-9.
15 Feuerbach, op. cit., pp. 204-5.
16 Quoted in A. R. Burn, *Pelican History of Greece*, p. 132.
17 G. Childe, *Man Makes Himself*, pp 107-8.
18 Trotsky, *In Defence of Marxism*, p. 66.
19 Marx, *Capital, Vol. 1*, p. 19.
20 David Bohm, *Causality and Chance in Modern Physics*, p. 1.
21 R. P. Feynman, *Lectures on Physics*, chapter 1, p. 8.
22 Aristotle, op. cit., p. 9.
23 Engels, *Dialectics of Nature*, p. 92.
24 Trotsky, op. cit., pp. 106-7.
25 M. Waldrop, *Complexity*, p. 82.
26 Engels, *Dialectics of Nature*, pp. 90-1.
27 Engels, *Anti-Dühring*, p. 162.
28 J. Gleick, *Chaos, Making a New Science*, p. 127.
29 M. Waldrop, op. cit., p. 65.
30 D. Bohm, op. cit., p. x.
31 Engels, *Anti-Dühring*, p. 163.
32 I. Stewart, *Does God Play Dice?* p. 22.
33 Feynman, op. cit., chapter 2, p. 5.
34 Engels, *Dialectics of Nature*, pp. 345-6.
35 Hegel, *Science of Logic, Vol. 1*, p. 258.
36 B. Hoffmann, *The Strange Story of the Quantum*, p. 159.
37 Engels, *Dialectics of Nature*, p. 96.
38 Ibid., pp. 95-6 and p. 110.
39 Ibid., p. 108 and p. 107.
40 Hegel, *Phenomenology of Mind*, p. 68.
41 Lenin, *Collected Works, Vol. 38*, p. 319; henceforth referred to as *LCW*.
42 Marx and Engels, *Selected Works, Vol. 1*, p. 502; henceforth referred to as *MESW*.

43 Marx, *Capital, Vol. 1*, pp. 19-20.
44 Quoted in A. A. Luce, Logic, p. 8.
45 *LCW*, Vol. 38, p. 171.
46 M. Donaldson, *Making Sense*, pp. 98-9.
47 M. Donaldson, *Children's Minds*, p. 76.
48 Trotsky, *Writings, 1939-40*, p. 400.
49 A. A. Luce, op. cit., p. 83.
50 Kant, *Critique of Pure Reason*, p. 99, footnote.
51 Engels, *The Dialectics of Nature*, pp. 64-5.
52 Feynman, op. cit., chapter 1, p. 2.
53 Kant, *Prolegomena zu einer jeden künftigen Metaphysik*, quoted in E. V. Ilyenkov, *Dialectical Logic*, p. 90.
54 Engels, *Anti-Dühring*, p. 43.
55 Trotsky, *Writings, 1939-40*, pp. 399 and 400.
56 Trotsky, *In Defence of Marxism*, p. 65, our emphasis.
57 Ibid., pp. 63-6.
58 Engels, *Anti-Dühring*, pp. 26-7.
59 Trotsky, *Writings, 1939-40*, pp. 401 and 403.
60 Cohen and Nagel, An *Introduction to Logic and the Scientific Method*, p. vii.

PART TWO: TIME, SPACE AND MOTION

5.   REVOLUTION IN PHYSICS

Two thousand years ago, it was thought that the laws of the universe were completely covered by Euclid's geometry. There was nothing more to be said. This is the illusion of every period. For a long time after Newton's death, scientists thought that he had said the last word about the laws of nature. Laplace lamented that there was only one universe, and that Newton had had the good fortune to discover all its laws. For two hundred years, Newton's particle theory of light was generally accepted, as against the theory, advocated by the Dutch physicist Huygens, that light was a wave. Then the particle theory was negated by the Frenchman, A. J. Fresnel, whose wave theory was experimentally confirmed by J. B. L. Foucault. Newton had predicted that light, which travels at 186,000 miles per second in empty space, should travel faster in water. The supporters of the wave theory predicted a lower speed, and were shown to be correct.

The great breakthrough for wave theory, however, was accomplished by the outstanding Scottish scientist James Clerk Maxwell, in the latter half of the 19th century. Maxwell based himself in the first instance on the experimental work of Michael Faraday, who discovered electromagnetic induction, and investigated the properties of the magnet, with its two poles, north and south, involving invisible forces stretching to the ends of the earth. Maxwell gave these empirical discoveries a universal form by translating them into mathematics. His work led to the discovery of the field, on which Einstein later based his general theory of relativity. One generation stands on the shoulders of its predecessors, both negating and preserving earlier discoveries, continually deepening them, and giving them a more general form and content.

Seven years after Maxwell's death, Hertz first detected the electromagnetic waves predicted by Maxwell. The particle theory, which had held sway ever since Newton, appeared to be annihilated by Maxwell's electromagnetics. Once again, scientists believed themselves in possession of a theory which could explain everything. There were just a few questions to be cleared up, and we would really know all there was to know about the workings of the universe. Of course, there were a few discrepancies which were troublesome, but they appeared to be small details which could safely be ignored. However, within a few decades, these "*minor*" discrepancies proved sufficient to overthrow the entire edifice and effect a veritable scientific revolution.

Waves or Particles?

Everyone knows what a wave is. It is a common feature associated with water. Just as waves can be caused by a duck moving over the surface of a pond, so a charged particle, an electron for example, can cause an electromagnetic wave, when it moves through space. The oscillatory motions of the electron disturbs the electric and magnetic fields, causing waves to spread out continuously, like the ripples on the pond. Of course, the analogy is only approximate. There is a fundamental difference between a wave on water and an electromagnetic wave. The latter does not require a continuous medium through which to travel, like water. An electromagnetic oscillation is a periodical disturbance that propagates itself through the electrical structure of matter. However, the comparison may help to make the idea clearer.

The fact that we cannot see these waves does not mean that their presence cannot be detected even in everyday life. We have direct experience of light-waves and radio-waves, and even X-rays. The only differences between them are their frequency. We know that a wave on water will cause a floating object to bob up and down faster or slower, depending on the intensity of the wave — the ripples caused by the duck, as compared to those provoked by a speed-boat. Similarly, the oscillations of the electrons will be proportionate to the intensity of the light-wave.

The equations of Maxwell, backed up by the experiments of Hertz and others, provided powerful evidence to support the theory that light consisted of waves, which were electromagnetic in character. However, at the turn of the century, evidence was accumulating which suggested that this theory was wrong. In 1900 Max Planck had shown that the classical wave theory made predictions which were not verified in practice. He suggested that light came in discrete particles or *"packets"* (quanta). The situation was complicated by the fact that different experiments proved different things. It could be shown that an electron was a particle by letting it strike a fluorescent screen and observing the resulting scintillations; or by watching the tracks made by electrons in a cloud chamber; or by the tiny spot that appeared on a developed photographer's plate. On the other hand, if two holes are made in a screen, and electrons were allowed to flood in from a single source, they caused an interference pattern, which indicated the presence of a wave.

The most peculiar result of all, however, was obtained in the celebrated two-slot experiment, in which a single electron is fired at a screen containing two slots and a photographer's plate behind it. Which of the two holes did the electron pass through? The interference pattern on the plate is quite clearly a two-hole pattern. This proves that the electron must have gone through both holes, and then set up an interference pattern. This is against all the laws of common sense, but it has been shown to be irrefutable. The electron behaves both like a particle and a wave. It is in two (or more than two) places at once, and in several states of motion at once!

"Let us not imagine," comments Banesh Hoffmann, *"that scientists accepted these new ideas with cries of joy. They fought them and resisted them as much as they could, inventing all sorts of traps and alternative hypotheses in vain attempts to escape them. But the glaring paradoxes were there as early as 1905 in the case of light, and even earlier, and no one had the courage or wit to resolve them until the advent of the new quantum mechanics. The new ideas are so difficult to accept because we still instinctively strive to picture them in terms of the old-fashioned particle, despite Heisenberg's indeterminacy principle. We still shrink from visualizing an electron as something which, having motion, may have no position, and having position, may have no such thing as motion or rest."* [1]

Here we see the negation of the negation at work. At first sight, we seem to have come full circle. Newton's particle theory of light was negated by Maxwell's wave theory. This, in turn, was negated by the new particle theory, advocated by Planck and Einstein. Yet this does not mean going back to the old Newtonian theory, but a qualitative leap forward, involving a genuine revolution in science. All of science had to be overhauled, including Newton's law of gravitation.

This revolution did not invalidate Maxwell's equations, which still remain valid for a vast field of operations. It merely showed that, beyond certain limits, the ideas of classical physics no longer apply. The phenomena of the world of subatomic particles cannot be understood by the methods of classical mechanics. Here the ideas of quantum mechanics and relativity come into play.

For most of the present century, physics has been dominated by the theory of relativity and quantum mechanics which, in the beginning, were rejected out of hand by the scientific establishment, which clung tenaciously to the old views. There is an important lesson here. Any attempt to impose a "final solution" to our view of the universe is doomed to fail.

Quantum Mechanics

The development of quantum physics represented a giant step forward in science, a decisive break with the old stultifying mechanical determinism of "classical" physics. (The "metaphysical" method, as Engels would have called it.) Instead, we have a much more flexible, dynamic — in a word dialectical — view of nature. Beginning with Planck's discovery of the existence of the quantum, which at first appeared to be a tiny detail, almost an anecdote, the face of physics was transformed. Here was a new science which could explain the phenomenon of radioactive transformation and analyze in great detail the complex data of spectroscopy. It directly led to the establishment of a new science — theoretical chemistry, capable of solving previously insoluble questions. In general, a whole series of theoretical difficulties were eliminated, once the new standpoint was accepted. The new physics revealed the staggering forces locked up within the atomic nucleus. This led directly to the exploitation of nuclear energy — the path to the potential destruction of life on earth — or the vista of undreamed of and limitless abundance and social progress through the peaceful use of nuclear fusion. Einstein's theory of relativity explains that mass and energy are equivalents. If the mass of an object is known, by multiplying it by the square of the speed of light, it becomes energy.

Einstein showed that light, hitherto thought of as a wave, behaved like a particle. Light, in other words, is just another form of matter. This was proved in 1919, when it was shown that light bends under the force of gravity. Louis de Broglie later pointed out that matter, which was thought to consist of particles, partakes of the nature of waves. The division between matter and energy was abolished once and for all. Matter and energy are...the same. Here was a mighty advance for science. And from the standpoint of dialectical materialism matter and energy are the same. Engels described energy (*"motion"*) as *"the mode of existence, the inherent attribute, of matter."* [2]

The argument which dominated particle physics for many years, whether subatomic particles like photons and electrons were particles or waves was finally resolved by quantum mechanics which asserts that subatomic particles can, and do, behave both like a particle and like a wave. Like a wave, light produces interferences, yet a photon of light also bounces off all electrons, like a particle. This goes against the laws of formal logic. How can *"common sense"* accept that an electron can be in two places at the same time? Or even move, at incredible speeds, simultaneously, in different directions? For light to behave both as a wave and as a particle was seen as an intolerable contradiction. The attempts to explain the contradictory phenomena of the subatomic world in terms of formal logic leads to the abandonment of rational thinking all together. In his conclusion to a work dealing with the quantum revolution, Banesh Hoffmann is capable of writing:

"How much more, then, shall we marvel at the wondrous powers of God who created the heaven and the earth from a primal essence of such exquisite subtlety that with it he could fashion brains and minds afire with the divine gift of clairvoyance to penetrate his mysteries. If the mind of a mere Bohr or Einstein astound us with its power, how may we begin to extol the glory of God who created them?" [3]

Unfortunately, this is not an isolated example. A great part of modern literature about science, including a lot written by scientists themselves, is thoroughly impregnated with such mystical, religious or quasi-religious notions. This is a direct result of the idealist philosophy which a great many scientists, consciously or unconsciously, have adopted.

The laws of quantum mechanics fly in the face of "common sense" (i.e., formal logic), but are in perfect consonance with dialectical materialism. Take, for example, the conception of a point. All traditional geometry is derived from a point, which subsequently becomes a line, a plane, a cube, etc. Yet close observation reveals that the point does not exist.

The point is conceived as the smallest expression of space, something which has no dimension. In reality, such a point consists of atoms — electrons, nuclei, photons, and even smaller particles. Ultimately, it disappears in a restless flux of swirling quantum waves. And there is no end to this process. No fixed *"point"* at all. That is the final answer to the idealists who seek to find perfect *"forms"* which allegedly lie *"beyond"* observable material reality. The only *"ultimate reality"* is the infinite, eternal, ever-changing material universe, which is far more wonderful in its endless variety of form and processes than the most fabulous adventures of science fiction. Instead of a fixed location — a "point" — we have a process, a never-ending flux. All attempts to impose a limit on this, in the form of a beginning or an end, will inevitably fail.

Disappearance of Matter?

Long before the discovery of relativity, science had discovered two fundamental principles — the conservation of energy and the conservation of mass. The first of these was worked out by Leibniz in the 17th century, and subsequently developed in the 19th century as a corollary of a principle of mechanics. Long before that, early man discovered in practice the principle of the equivalence of work and heat, when he made fire by means of friction, thus translating a given amount of energy (work) into heat. At the beginning of this century, it was discovered that mass is merely one of the forms of energy. A particle of matter is nothing more than energy, highly concentrated and localized. The amount of energy concentrated in a particle is proportional to its mass, and the total amount of energy always remains the same. The loss of one kind of energy is compensated for by the gain of another kind of energy. While constantly changing its form, nevertheless, energy always remains the same.

The revolution effected by Einstein was to show that mass itself contains a staggering amount of energy. The equivalence of mass and energy is expressed by the formula $E = mc^2$ in which c represents the velocity of light (about 186,000 miles per second), E is the energy that is contained in the stationary body, and m is its mass. The energy contained in the mass m is equal to this mass, multiplied by the square of the tremendous speed of light. Mass is therefore an immensely concentrated form of energy, the power of which may be conveyed by the fact that the energy released by an atomic explosion is less than one tenth of one per cent of the mass converted into energy. Normally this vast amount of energy locked up in matter is not manifested, and therefore passes unnoticed. But if the processes within the nucleus reach a critical point, part of the energy is released, as kinetic energy.

Since mass is only one of the forms of energy, matter and energy can neither be created nor destroyed. The forms of energy, on the other hand, are extremely diverse. For example, when protons in the sun unite to form helium nuclei, nuclear energy is released. This may first appear as the kinetic energy of motion of nuclei, contributing to the heat energy from the sun. Part of this energy is emitted from the sun in the form of photons, containing particles of electro-magnetic energy. The latter, in turn, is transformed by the process of photosyn-thesis into the stored chemical energy in plants, which, in turn, is acquired by man by eating the plants, or animals which have fed upon the plants, to provide the warmth and energy for muscles, blood circulation, brain, etc.

The laws of classical physics in general cannot be applied to processes at the subatomic level. However, there is one law which knows no exception in nature — the law of the conservation of energy. Physicists know that neither a positive nor a negative charge can be created out of nothing. This fact is expressed by the law of the conservation of electric charge. Thus, in the process of producing a beta particle, the disappearance of the neutron (which has no charge) gives rise to a pair of particles with opposed charges — a positively-charged proton and a negatively-charged electron. Taken together, the two new particles have a combined electrical charge equal to zero.

If we take the opposite process, when a proton emits a positron and changes into a neutron, the charge of the original particle (the proton) is positive, and the resulting pair of particles (the neutron and positron), taken together, are positively charged. In all these myriad changes, the law of the conservation of electrical charge is strictly maintained, as are all the other conservation laws. Not even the tiniest fraction of energy is created or destroyed. Nor will such a phenomenon ever occur.

When an electron and its anti-particle, the positron, destroy themselves, their mass "*disappears*," that is to say, it is transformed into two light-particles (photons) which fly apart in opposite directions. However, these have the same total energy as the particles from which they emerged. Mass-energy, linear momentum and electric charge are all preserved. This phenomenon has nothing in common with disappearance in the sense of annihilation. Dialectically, the electron and positron are negated and preserved at the same time. Matter and energy (which is merely two ways of saying the same thing) can neither be created nor destroyed, only transformed.

From the standpoint of dialectical materialism, matter is the objective reality given to us in sense-perception. That includes not just "*solid*" objects, but also light. Photons are just as much matter as electrons or positrons. Mass is constantly being changed into energy (including light — photons) and energy into mass. The "*annihilation*" of a positron and an electron produces a pair of photons, but we also see the opposite process: when two photons meet, an electron and a positron can be produced, provided that the photons possess sufficient energy. This is sometimes presented as the creation of matter "*from nothing*." It is no such thing. What we see here is neither the destruction nor the creation of anything, but the continuous transformation of matter into energy, and vice versa. When a photon hits an atom, it ceases to exist as a photon. It vanishes, but causes a change in the atom — an electron jumps from a one orbit to another of higher energy. Here too, the opposite process occurs. When an electron jumps to an orbit of lower energy, a photon emerges.

The process of continual change which characterizes the world at the subatomic level is a striking confirmation of the fact that dialectics is not just a subjective invention of the mind, but actually corresponds to objective processes taking place in nature. This process has gone on uninterruptedly for all eternity. It is a concrete demonstration of the indestructibility of matter — precisely the opposite of what it was meant to prove.

"Bricks of Matter"?

For centuries, scientists have tried in vain to find the "*bricks of matter*" — the ultimate, smallest particle. A hundred years ago, they thought they had found it in the atom (which, in Greek, signifies *"that which cannot be divided"*). The discovery of subatomic particles led physics to probe deeper into the structure of matter. By 1928 scientists imagined that they had discovered the smallest particles — protons, electrons and photons. All the material world was supposed to be made up of these three. Subsequently, this was shattered by the discovery of the neutron, the positron, the deuteron, then a host of other particles, ever smaller, with an increasingly fleeting existence — neutrinos, pi-mesons, mu-mesons, k-mesons, and many others. The life-span of some of these particles is so evanescent — maybe a billionth of a second — that they have been described as *"virtual particles"* — something utterly unthinkable in the pre-quantum era.

The tauon lasts only for a trillionth of a second, before breaking down into a muon, and then to an electron. The neutral pion is even more fleeting, breaking down in less than one quadrillionth of a second to form a pair of gamma rays. However, these gammas live to a ripe old age, compared to others which have a life of only one hundredth of a microsecond. Some, like the neutral sigma particle, break down after a hundred trillionth of a second. In the 1960s, even this was overtaken by the discovery of particles so evanescent that their existence could only be determined from the necessity of explaining their breakdown products. The half-lives of these particles are in the region of a few trillionths of a second. These are known as resonance particles. And even this was not the end of the story.

Over a hundred and fifty new particles were later discovered, which have been called hadrons. The situation was becoming extremely confused. An American physicist, Dr. Murray Gell-Mann, in an attempt to explain the structure of subatomic particles, postulated still other, more basic particles, the quarks, which were yet again heralded as the "ultimate building-blocks of matter." Gell-Mann theorized that there were six different kinds of quarks and that the quark family was parallel to a six member family of lighter particles known as leptons. All matter was now supposed to consist of these twelve particles. Even these, the most basic forms of matter so far known to science, still possess the same contradictory qualities we observe throughout nature, in accordance with the dialectical law of the unity of opposites. Quarks also exist in pairs, and possess a positive and negative charge, although it is, unusually, expressed in fractions.

Despite the fact that experience has demonstrated that there is no limit to matter, scientists still persist in the vain search for the "bricks of matter." It is true that such expressions are the sensational inventions of journalists and some scientists with an over-developed flare for self-promotion, and that the search for ever smaller and fundamental particles is undoubtedly a bona-fide scientific activity, which serves to deepen our knowledge of the workings of nature.

Nevertheless, one certainly gets the impression that at least some of them really do believe that it is possible to reach a kind of ultimate level of reality, beyond which there is nothing left to discover, at least at the subatomic level.

The quark is supposed to be the last of twelve subatomic *"building blocks"* which are said to make up all matter. *"The exciting thing is that this is the final piece of matter as we know it, as predicted by cosmology and the Standard Model of particle physics, Dr. David Schramm was reported as saying, 'It is the final piece of that puzzle.'"* [4] So the quark is the *"ultimate particle."* It is said to be fundamental and structureless. But similar claims were made in the past for the atom, then the proton, and so on and so forth. And in the same way, we can confidently predict the discovery of still more "fundamental" forms of matter in the future. The fact that the present state of our knowledge and technology does not permit us to determine the properties of the quark does not entitle us to affirm that it has no structure. The properties of the quark still await analysis, and there is no reason to suppose that this will not be achieved, pointing the way to a still deeper probing of the endless properties of matter. This is the way science has always advanced. The supposedly unbreachable barriers to knowledge erected by one generation are overturned by the next, and so on down the ages. The whole of previous experience gives us every reason to believe that this dialectical process of the advance of human knowledge is as endless as the infinite universe itself.

6.   UNCERTAINTY AND IDEALISM

The Uncertainty Principle

The real death knell for Newtonian mechanics as an universal theory was sounded by Einstein, Schrödinger, Heisenberg and other scientists that stood at the cradle of quantum mechanics in the early 20th century. The behavior of *"elementary particles"* could not be explained by classical mechanics. A new mathematics had to be developed.

In this mathematics there are concepts like a *"phase-space"* wherein a system is defined as a point which has its degrees of freedom as coordinates, and "operators," magnitudes that are incompatible with algebraic magnitudes in the sense that they are more similar to operations than to magnitudes themselves (in fact they express relations instead of fixed properties) play a significant role. Probability also plays an important role, but in the sense of *"intrinsic probability"*: it is one of the essential characteristics of quantum mechanics. In fact quantum mechanic systems must be interpreted as the superposition of all the possible pathways they can follow.

Quantum particles can only be defined as a set of internal relationships between their *"actual"* and its *"virtual"* state. In that sense they are purely dialectical. Measuring those particles in one way or another leads only to the revealing of the "*actual*" state which is only one aspect of the whole (this paradox is popularly explained by the tale of "*Schrödinger's cat*"). It is called the "collapse of the wave function," and is expressed by the uncertainty principle of Heisenberg. This entirely new way of looking toward physical reality, which is expressed by quantum mechanics, was kept "in quarantine" for long time by the rest of the scientific disciplines. It was seen as an exceptional kind of mechanics, only to be used in describing the behavior of elementary particles, the exception to the rule of classic mechanics, without any importance whatsoever.

In place of the old certainties, uncertainty now reigned. The apparently random movements of subatomic particles, with their unimaginable velocities, could not be expressed in terms of the old mechanics. When a science reaches a blind alley, when it is no longer able to explain the facts, the ground is prepared for a revolution, and the emergence of a new science. However, the new science, in its initial form, is not yet completely developed. Only over a period does it emerge in its final and complete form. A degree of improvisation, of uncertainty, of varying and often contradictory interpretations, is virtually inevitable at first.

In recent decades a debate has opened up between the so-called "*stochastic*" ("*random*") interpretation of nature and determinism. The fundamental problem is that necessity and chance are here treated as absolute opposites, mutually exclusive contraries. In this way, we arrive at two opposing views, neither of which is adequate to explain the contradictory and complex workings of nature.

Werner Heisenberg, a German physicist, developed his own peculiar version of quantum mechanics. In 1932, he received the Nobel Prize for physics for his system of matrix mechanics, which described the energy levels of orbits of electrons purely in terms of numbers, without any recourse to pictures. In this way, he hoped to get round the problems caused by the contradiction between "particles" and "waves" by abandoning any attempt to visualize the phenomenon, and treating it in a purely mathematical abstraction. Erwin Schrödinger's wave mechanics covered exactly the same ground as Heisenberg's matrix mechanics without any need to retreat into the realms of absolute mathematical abstraction. Most physicists preferred Schrödinger's approach, which seemed far less abstract, and they were not wrong. In 1944, John van Neumann, the Hungarian-American mathematician, demonstrated that wave mechanics and matrix mechanics were mathematically equivalent, and could achieve exactly the same results.

Heisenberg achieved some important advances in quantum mechanics. However, permeating his whole approach was the determination to inflict his peculiar brand of philosophical idealism upon the new science. From this arose the so-called "Copenhagen interpretation" of quantum mechanics. This was really a variety of subjective idealism, thinly disguised as a school of scientific thought. "Werner Heisenberg," wrote Isaac Asimov, "proceeded to raise a profound question that projected particles, and physics itself, almost into a realm of the unknowable." [5] That is the correct word to use. We are not dealing here with the unknown. That is always present in science. The whole history of science is the advance from the unknown to the known, from ignorance to knowledge. But a serious difficulty arises when people confuse the unknown with the unknowable. There is a fundamental difference between the words "we do not know" and *"we cannot know."* Science sets out from the basic notion that the objective world exists and can be known to us.

However, in the whole history of philosophy there have been repeated attempts to place a limit upon human cognition, to assert that there are certain things which "we cannot know," for this reason or that. Thus Kant claimed that we could only know appearances, but not Things-in-Themselves. In this, he was following in the footsteps of the skepticism of Hume, the subjective idealism of Berkeley and the sophists: that we cannot know the world.

In 1927, Werner Heisenberg advanced his celebrated *"uncertainty principle,"* according to which it is impossible to determine, with the desired accuracy, both the position and velocity of a particle simultaneously. The more certain a particle's position, the more uncertain its momentum, and vice versa. (This also applies to other specified pairs of properties.) The difficulty in establishing precisely the position and velocity of a particle which is moving at 5,000 miles per second in different directions is self-evident. However, to deduce from this that cause and effect (causality) in general does not exist is an entirely false proposition.

How can we decide on the position of an electron? he asked. By looking at it. But if we use a powerful microscope, it would mean striking it with a particle of light, a photon. Because light behaves like a particle, it will inevitably disturb the momentum of the observed particle. Therefore, we change it by the very act of observation. The disturbance will be unpredictable and uncontrollable, since (at least from the existing quantum theory) there is no way of knowing or controlling beforehand the precise angle with which the light quantum will be scattered into the lens. Because an accurate determination of the position requires the use of light of short wave-length, a large but unpredictable and uncontrollable momentum is transferred to the electron. On the other hand, an accurate determination of the momentum requires the use of light quanta of very low momentum (and therefore of long wave-length), which means a large angle of diffraction, and hence a poor definition of the position. The more accurately the position is defined, the less accurate the momentum can be defined, and vice versa.

So can we get round this problem if we develop new kinds of electron microscopes? Not according to Heisenberg's theory. Since all energy comes in quanta, and all matter has the property of acting both as a wave and a particle, any type of apparatus we use will be governed by this principle of uncertainty (or indeterminacy). Indeed, the term uncertainty principle is inexact, because what is asserted here is not just that we cannot be certain, because of problems of measurement. The theory implies that all forms of matter are indeterminate by their very nature. As David Bohm says in his book *Causality and Chance in Modern Physics*:

> *Thus the renunciation of causality in the usual interpretation of the quantum theory is not to be regarded as merely the result of our inability to measure the precise values of the variables that would enter into the expression of causal laws at the atomic level, but, rather, it should be regarded as a reflection of the fact that no such laws exist.*

Instead of seeing it as a special aspect of quantum theory at a particular stage in its development, Heisenberg postulated indeterminacy as a fundamental and universal law of nature, and assumed that all other laws of nature would have to be consistent with it. This is completely different to the approach of science in the past when it was confronted with problems related to irregular fluctuations and random movement. No-one imagines it is possible to determine the exact motion of an individual molecule in a gas, or predict all the details of a specific car accident. But never before has a serious attempt been made to derive from such facts the non-existence of causality in general.

Yet this is precisely the conclusion we are invited to draw from the principle of indeterminacy. Scientists and idealist philosophers have gone on to argue that causality in general does not exist. That is to say, that there is no cause and effect. Nature thus appears as an entirely causeless, random affair. The entire universe is unpredictable. *"We cannot be certain"* of anything. *"Instead, it is assumed that in any particular experiment, the precise result that will be obtained is completely arbitrary in the sense that it has no relationship whatever to anything else that exists in the world or that ever has existed."* [6]

This position is the complete negation, not only of science, but of rational thought in general. If there is no cause and effect, not only is it impossible to predict anything; it is impossible to explain anything. We can only limit ourselves to describe what is. In fact, not even that, since we cannot even be certain that anything exists outside ourselves and our own senses. This brings us right back to the philosophy of subjective idealism. It reminds us of the argument of the sophist philosophers of ancient Greece: *"I cannot know anything about the world. If I can know something, I cannot understand it. If I can understand it, I cannot express it."*

What the *"indeterminacy principle"* really represents is the highly elusive character of the movement of subatomic particles, which are not susceptible to the kind of simplistic equations and measurements of classical mechanics. There is no doubt about Heisenberg's contribution to physics. What is in question is the philosophical conclusions which he drew from quantum mechanics. The fact that we cannot measure exactly the position and momentum of an electron does not imply in the slightest that there is a lack of objectivity here. The subjective way of thinking permeates the so-called Copenhagen school of quantum mechanics. Niels Bohr went so far as to state that "it is wrong to think that the task of physics is to find out how nature is. Physics concerns what we can say about nature."

The physicist John Wheeler maintains that *"no phenomenon is a real phenomenon until it is an observed phenomenon."* And Max Born spells out the same subjectivist philosophy with absolute clarity: "The generation to which Einstein, Bohr, and I belong was taught that there exists an objective physical world, which unfolds itself according to immutable laws independent of us; we are watching this process as the audience watches a play in a theatre. Einstein still believes that this should be the relation between the scientific observer and his subject." [7]

What we have here is not a scientific evaluation, but a philosophical opinion reflecting a definite world outlook — that of subjective idealism, which permeates the entire Copenhagen interpretation of quantum theory. A number of eminent scientists, to their credit, made a stand against this subjectivism, which runs contrary to the whole outlook and method of science. Among these were Einstein, Max Planck, Louis de Broglie and Erwin Schrödinger, all of whom played a role in developing the new physics at the very least as important as Heisenberg.

Objectivity Versus Subjectivism

There is not the slightest doubt that Heisenberg's interpretation of quantum physics was heavily influenced by his philosophical views. Even as a student, Heisenberg was a conscious idealist, who admits being greatly impressed by Plato's *Timaeus* (where Plato's idealism is expressed in the most obscurantist way), while fighting in the ranks of the reactionary *Freikorps* against the German workers in 1919. Subsequently he stated that he was "much more interested in the underlying philosophical ideas than in the rest," and that it was necessary *"to get away from the idea of objective processes in time and space."* In other words, Heisenberg's philosophical interpretation of quantum physics was very far from being the objective result of scientific experiment. It was clearly linked to idealist philosophy, which he consciously applied to physics, and which determined his outlook.

Such a philosophy is at odds not only with science, but the whole of human experience. Not only does it lack any scientific content, but it turns out to be perfectly useless in practice. Scientists who, as a rule, like to steer clear of philosophical speculation, make a polite nod in the direction of Heisenberg, and simply get on with the job of investigating the laws of nature, taking for granted not only that it exists, but that it functions according to definite laws, including those of cause and effect, and that, with a bit of effort, can be perfectly well understood, and even predicted by men and women. The reactionary consequences of this subjective idealism are shown by Heisenberg's own evolution. He justified his active collaboration with the Nazis on the grounds that "There are no general guidelines to which we can cling. We have to decide for ourselves, and cannot tell in advance if we are doing right or wrong." [8]

Erwin Schrödinger did not deny the existence of random phenomena in nature in general or in quantum mechanics. He specifically mentions the example of the random combining of DNA molecules at the moment of conception of a child, in which the quantum features of the chemical bond play a role. However, he objected to the standard Copenhagen interpretation about the implications of the "two-hole" experiment; that Max Born's waves of probability meant that we had to renounce the objectivity of the world, the idea that the world exists independently of our observing it.

Schrödinger ridiculed the assertion of Heisenberg and Bohr that, when an electron or photon is not being observed, it has "no position" and only materializes at a given point as a result of the observation. To counter it, he devised a famous "thought experiment." Take a cat and put it in a box with a vial of cyanide, he said. When a Geiger counter detects the decay of an atom, the vial is broken. According to Heisenberg, the atom does not "know" it has decayed until someone measures it. In this case, therefore, until someone opens the box and looks in, according to the idealists, the cat is neither dead nor alive! By this anecdote, Schrödinger meant to highlight the absurd contradictions caused by the acceptance of Heisenberg's subjective idealist interpretation of quantum physics. The processes of nature take place objectively, irrespective of whether human beings are around to observe them or not.

According to the Copenhagen interpretation, reality only comes into being when we observe it. Otherwise, it exists in a kind of limbo, or "probability wave superposition state," like our live-and-dead cat. The Copenhagen interpretation draws a sharp line of distinction between the observer and the observed. Some physicists take the view, following the Copenhagen interpretation, that consciousness must exist, but the idea of material reality without consciousness is unthinkable. This is precisely the standpoint of subjective idealism which Lenin comprehensively answered in his book *Materialism and Empirio-criticism.*

Dialectical materialism sets out from the objectivity of the material universe, which is given to us through sense perception. "I interpret the world through my senses." That is self-evident. But the world exists independently of my senses. That is also self-evident, one might think, but not for modern bourgeois philosophy! One of the main strands of 20th century philosophy is logical positivism, which precisely denies the objectivity of the material world. More correctly, it considers that the very question of whether the world exists or not to be irrelevant and *"metaphysical."* The standpoint of subjective idealism has been completely undermined by the discoveries of 20th century science. The act of observation means that our eyes are receiving energy from an external source in the form of light waves (photons). This was clearly explained by Lenin in 1908-9:

If color is a sensation only depending upon the retina (as natural science compels you to admit), then light rays, falling upon the retina, produce the sensation of color. This means that outside us, independently of us and of our minds, there exists a movement of matter, let us say of ether waves of a definite length and of a definite velocity, which, acting upon the retina, produce the sensation of color. This is precisely how natural science regards it. It explains the sensations of various colors by the various lengths of light waves existing outside the human retina, outside man and independently of him. This is materialism: matter acting upon our sense-organs produces sensation. Sensation depends on the brain, nerves, retina, etc., i.e., on matter organized in a definite way. The existence of matter does not depend on sensation. Matter is primary. Sensation, thought, consciousness are the supreme product of matter organized in a particular way. Such are the views of materialism in general, and of Marx and Engels in particular. [9]

The subjective idealist nature of Heisenberg's method is quite explicit:

Our actual situation in research work in atomic physics is usually this: we wish to understand a certain phenomenon, we wish to recognize how this phenomenon follows from the general laws of nature. Therefore, that part of matter or radiation which takes part in the phenomenon is the natural 'object' in the theoretical treatment and should be separated in this respect from the tools used to study the phenomenon. This again emphasizes a subjective element in the description of atomic events, since the measuring device has been constructed by the observer, and we have to remember that what we observe is not nature in itself but nature exposed to our method of questioning. Our scientific work in physics consists in asking questions about nature in the language that we possess and trying to get an answer from experiment by the means that are at our disposal. [10]

Kant erected an impenetrable barrier between the world of appearances and reality *"in itself."* Here Heisenberg goes one better. He not only speaks about "nature in itself," but even maintains that we cannot really know that part of nature which can be observed, since we change it by the very act of observing it. In this way, Heisenberg seeks to abolish the criterion of scientific objectivity altogether. Unfortunately, many scientists who would indignantly deny the charge of mysticism have uncritically assimilated Heisenberg's philosophical ideas, merely because they are unwilling to accept the necessity for a consistently materialist philosophical approach to nature.

The whole point is that the laws of formal logic break down beyond certain limits. This most certainly applies to the phenomena of the subatomic world, where the laws of identity, contradiction and the excluded middle cannot be applied. Heisenberg defends the standpoint of formal logic and idealism, and therefore, inevitably arrives at the conclusion that the contradictory phenomena at the subatomic level cannot be comprehended by human thought at all. The contradiction, however, is not in the observed phenomena at the subatomic level, but in the hopelessly antiquated and inadequate mental schema of formal logic. The so-called "paradoxes of quantum mechanics" are precisely this. Heisenberg cannot accept the existence of dialectical contradictions, and therefore prefers to revert to philosophical mysticism — *"we cannot know,"* and all the rest of it.

We find ourselves here in the presence of a kind of philosophical conjuring trick. The first step is to confuse the concept of causality with the old mechanical determinism represented by people like Laplace. These limitations were explained and criticized by Engels in the *Dialectics of Nature*. The discoveries of quantum mechanics finally destroyed the old mechanical determinism. The kind of predictions made by quantum mechanics are somewhat different from those of classical mechanics. Yet quantum mechanics still makes predictions, and obtains precise results from them.

Causality and Chance

One of the problems faced by the student of philosophy or science is when a particular terminology is used that is frequently at variance with everyday language. One of the fundamental problems in the history of philosophy is the relationship between freedom and necessity, a complex question, which is not made any easier when it emerges in different disguises — causality and chance, necessity and accident, determinism and indeterminism, etc.

We all know from everyday experience what we mean by necessity. When we need to do something, it means that we have no choice. We cannot do otherwise. The dictionary defines necessity as a set of circumstances compelling something to be, or to be done, especially relating to a law of the universe, inseparable from, and directing, human life and action. The idea of physical necessity involves the notion of compulsion and constraint. It is conveyed by expressions like "to bow to necessity." It occurs in proverbs like *"necessity knows no law."*

In the philosophical sense, necessity is closely related to causality, the relation between cause and effect — a given action or event necessarily gives rise to a particular result. For example, if I stop breathing for an hour, I will die, or if I rub two sticks together, I will produce heat. This relation between cause and effect, which is confirmed by an infinite number of observations and practical experiences, plays a central role in science. By contrast, accident is regarded as an unexpected event, which occurs without apparent cause, as when we trip over a loose paving stone, or drop a cup in the kitchen. In philosophy, however, accident is a property of a thing which is a merely contingent attribute, that is, something which is not part of its essential nature. An accident is something which does not exist of necessity, and which equally well could not have happened. Let us consider an example.

If I let this piece of paper go, it will normally fall to the floor, because of the law of gravity. That is an example of causation, of necessity. But if a sudden draught should cause the paper to blow away unexpectedly, that would be generally seen as chance. Necessity is therefore governed by law, and can be scientifically expressed and predicted. Things which happen of necessity are things which could not have happened otherwise. On the other hand, random events, contingencies, are events which might, or might not, happen; they are governed by no law which can be clearly expressed and are by their very nature, unpredictable.

Experience of life convinces us that both necessity and accident exist and play a role. The history of science and society shows exactly the same thing. The whole essence of the history of science is the search for the underlying patterns of nature. We learn early in life to distinguish between the essential and non-essential, the necessary and contingent. Even when we come across exceptional conditions which may seem "*irregular*" to us at a given stage of our knowledge, it often turns out that subsequent experience reveals a different kind of regularity, and still deeper causal relations, which were not immediately obvious.

The search for a rational insight and understanding of the world in which we live is intimately connected with the need to discover causality. A small child, in the process of learning about the world, will always ask "*why?*" — to the distraction of its parents, who are frequently at a loss for an answer. On the basis of observation and experience, we formulate a hypothesis as to what causes a given phenomenon. This is the basis of all rational understanding. As a rule, these hypotheses in turn give rise to predictions concerning things which have not yet been experienced. These may then be tested, either by observation or practice. This is not only a description of the history of science, but also of an important part of the mental development of every human being from early childhood on. It therefore covers intellectual development in the very broadest sense of the word, from the most basic learning processes of a child up to the most advanced study of the universe.

The existence of causality is shown by an immense number of observations. These enable us to make important predictions, not only in science, but in everyday life. Everyone knows that if water is heated to 100°C, it turns into steam. This is the basis not only for making a cup of tea, but for the industrial revolution, upon which the whole of modern society rests. Yet there are philosophers and scientists who seriously maintain that steam cannot said to be caused by heating water. The fact that we can make predictions about a vast number of events is itself proof that causality is not merely a convenient way of describing events, but, as David Bohm points out, an inherent and essential aspect of things. Indeed, it is impossible even to define the properties of things without resorting to causality. For example, when we say that something is red, we imply that it will react in a certain way when subjected to specified conditions — i.e., a red object is defined as one which when exposed to white light will reflect mostly red light. Similarly, the fact that water becomes steam when heated, and ice when cooled, is the expression of a qualitative causal relationship which is part of the essential properties of this liquid, without which it could not be water. The general mathematical laws of motion of moving bodies are likewise essential properties of these bodies, without which they could not be what they are. Such examples may be multiplied without limit. In order to understand why and how causality is so closely bound up with the essential properties of things, it is not enough to consider things statically and in isolation. It is necessary to consider things as they are, as they have been, and as they will necessarily become in the future — that is to say, to analyze things as processes.

In order to understand particular events, it is not necessary to specify all the causes. Indeed, this is not possible. The kind of absolute determinism put forward by Laplace was answered in advance by Spinoza in the following witty passage:

> For example, if a stone falls from a roof on the head of a passer-by and kills him, they will show by their method of argument that the stone was sent to fall and kill the man; for if it had not fallen on him for that end, by God's will, how could so many circumstances (for often very many circumstances concur at the same time) concur by chance? You will reply, perhaps: "The wind was blowing and the man had to pass that way, and hence it happened." But they will retort: "Why was the wind blowing at that time? And why was the man going that way at that time?" If again you reply: "The wind had then arisen on account of the agitation of the sea the day before, the previous weather having been calm, and the man was going that way at the invitation of a friend," they will again retort, for there is no end to their questioning: "Why was the sea agitated, and why was the man invited at that time?"

And thus they will pursue you from cause to cause until you are glad to take refuge in the will of God, that is, the asylum of ignorance. Thus again, when they see the human body they are amazed, and as they know not the cause of so much art, they conclude that it was not by mechanical art, but divine or supernatural art, and constructed in such a manner that one part does not injure another. And hence it comes about that someone who wishes to seek out the true causes of miracles, and to understand the things of nature like a man of learning, and not to stare at them in amazement like a fool, is widely deemed heretical and impious, and proclaimed such by those whom the mob adore as interpreters of nature and the Gods. For these know that once ignorance is laid aside, that wonderment which is their only means of arguing and of preserving their authority would be taken away.[11]

Mechanism

The attempt to eliminate all contingency from nature leads necessarily to a mechanistic viewpoint. In the mechanistic philosophy of the 18th century — represented in science by Newton, the bare idea of necessity was elevated to an absolute principle. It was seen as perfectly simple, free from all contradiction, and with no irregularities or cross-currents.

The idea of the universal lawfulness of nature is profoundly true, but a bare statement of lawfulness is insufficient. What is necessary is a concrete understanding of how the laws of nature actually operate. The mechanistic outlook necessarily developed a one-sided view of the phenomena of nature, reflecting the actual level of scientific development at the time. The highest achievement of this view was classical mechanics, which deals with relatively simple processes, cause and effect, understood as the simple external action of one solid body upon another, levers, equilibrium, mass, inertia, pushing, pressing, and the like. Important as these discoveries were, they were clearly insufficient to arrive at an accurate idea of the complex workings of nature. Later on, the discoveries of biology, particularly after the Darwinian revolution, made possible a different approach to scientific phenomena, in line with the more flexible and subtle processes of organic matter.

In classical Newtonian mechanics motion is treated as something simple. If we know at any given moment what different forces apply to a specific moving object, we can predict exactly how it will behave in the future. This leads to mechanistic determinism, the most prominent exponent of which was Pierre Simon de Laplace, the French 18th century mathematician, whose theory of the universe really is identical to the idea of predestination present in several religions, notably Calvinism.

In his *Philosophical Essays on Probabilities*, Laplace wrote:

An intellect which at any given moment knew all the forces that animate Nature and the mutual positions of the being that comprise it, if this intellect were vast enough to submit its data to analysis, could condense into a single formula the movement of the greatest bodies of the universe and that of the lightest atom: for such an intellect nothing could be uncertain; and the future just like the past would be present before our eyes. [12]

The difficulty arises from the mechanistic method inherited by 19th century physics from the 18th century. Here necessity and chance were regarded as fixed opposites, the one excluding the other. A thing or process was either accidental or necessary, but not both. This method was subjected to a searching analysis by Engels in *The Dialectics of Nature*, where he explains that the mechanistic determinism of Laplace inevitably led to fatalism and a mystical concept of nature:

And then it is declared that the necessary is the sole thing of scientific interest and that the accidental is a matter of indifference to science. That is to say: what can be brought under laws, hence what one knows, is interesting; what cannot be brought under laws, and therefore what one does not know, is a matter of indifference and can be ignored. Thereby all science comes to an end, for it has to investigate precisely that which we do not know. It means to say: what can be brought under general laws is regarded as necessary, and what cannot be so brought as accidental. Anyone can see that this is the same sort of science as that which proclaims natural what it can explain, and ascribes what it cannot explain to supernatural causes; whether I term the cause of the inexplicable chance, or whether I term it God, is a matter of complete indifference as far as the thing itself is concerned. Both are only equivalents for: I do not know, and therefore do not belong to science. The latter ceases where the requisite connection is wanting.

Engels points out that such mechanical determinism effectively reduces necessity to the level of chance. If every trifling occurrence is of the same order of importance and necessity as the universal law of gravity, then all fundamental laws are on the same level of triviality:

According to this conception only simple, direct necessity prevails in nature. That a particular pea-pod contains five peas and not four or six, that a particular dog's tail is five inches long and not a whit longer or shorter, that this year a particular clover flower was fertilized by a bee and another not, and indeed by precisely one particular bee and a particular time, that a particular windblown dandelion seed has sprouted and another not, that last night I was bitten by a flea at four o'clock in the morning, and not at three or five o'clock, and on the right shoulder and not on the left calf — these are all facts which have been produced by an irrevocable concatenation of cause and effect, by an unshatterable necessity of such a nature indeed that the gaseous sphere, from which the solar system was derived, was already so constituted that these events had to happen thus and not otherwise.

With this kind of necessity we likewise do not get away from the theological conception of nature. Whether with Augustine and Calvin we call it the eternal decree of God, or Kismet as the Turks do, or whether we call it necessity, is all pretty much the same for science. There is no question of tracing the chain of causation in any of these cases; so we are just as wise in one as in another, the so-called necessity remains an empty phrase, and with it — chance also remains what it was before. [13]

Laplace thought that if he could trace the causes of everything in the universe he could abolish contingency altogether. For a long time, it appeared that the workings of the entire universe could be reduced to a few relatively simple equations. One of the limitations of the classical mechanistic theory is that it assumes that there are no outside influences on the motion of particular bodies. In reality, however, everybody is influenced and determined by every other body. Nothing can be taken in isolation.

Nowadays the claims of Laplace seem extravagant and unreasonable. But then, similar extravagances are to be seen at every stage in the history of science, where each generation firmly believes itself to be in possession of the "ultimate truth." Nor is this entirely mistaken. The ideas of each generation are indeed the ultimate truth, for that period. But all that we are saying when we make such assertions is: *"This is as far as we have got in understanding Nature, with the information and technological capabilities we currently possess."* Therefore, it is not incorrect to claim that these truths are absolute for us at this moment in time since we can base ourselves on no others.

The 19th Century

Newton's classical mechanics in their time represented an enormous step forward in science. For the first time, Newton's laws of motion made possible precise quantitative predictions, which could be checked against the observed phenomena. However, precisely this precision leads to new problems when Laplace and others attempted to apply them to the universe as a whole. Laplace was convinced that Newton's laws were absolutely and universally valid. This was doubly incorrect. First of all, Newton's laws were not seen as approximations applicable in certain circumstances. Secondly Laplace did not consider the possibility that under different circumstances, in areas not yet studied in physics, these laws might need to be modified or extended. The mechanistic determinism of Laplace supposed that once the positions and velocities were known at any instant of time the future behavior of the whole universe would be determined for all time. According to this theory, all the rich diversity of things can be reduced to an absolute set of quantitative laws based on a few variables.

Classical mechanics as expressed in Newton's laws of motion deal with simple cause and effect, for example the isolated action of one body upon another. However, in practice, this is impossible, since no mechanical system is ever completely isolated. Outside influences inevitably destroy the isolated one-to-one character of the connection. Even if we could isolate the system, there will still be disturbances arisen from motions at the molecular level, and other disturbances at the even deeper level of quantum mechanics. As Bohm remarks: *"Thus, there is no real case known of a set of perfect one-to-one causal relationships that could in principle make possible predictions of unlimited precision, without the need to take into account qualitatively new sets of causal factors existing outside the system of interest or at other levels.*"[14]

Does this mean that prediction is impossible? Not at all. When we aim a gun at a certain point, the individual bullet will not land precisely at the point predicted by Newton's law of motion. However, a large number of shots fired will form a cluster in a small region near the point predicted. Thus, within a given range of error, which always exists, very precise predictions are possible. If we wanted to obtain unlimited precision in this instance, we would discover an ever increasing number of factors which influence the result — irregularities in the structure of the gun and bullet, tiny variations of temperature, pressure, humidity, air currents, and even the molecular motions of all these factors.

Some degree of approximation is necessary, which does not take into account the infinity of factors required for a perfectly precise prediction of a given result. This involves a necessary abstraction from reality, as in Newtonian mechanics. However, science continually proceeds, step by step, to discover ever deeper and more precise laws which enable us to gain a deeper understanding of the processes of nature, and thus make more accurate predictions. The abandonment of the old mechanical determinism of Newton and Laplace does not mean the abolition of causality, but a deeper understanding of the way in which causality actually works.

The first breaches in the wall of Newtonian science appeared in the second half of the 19th century, especially with Darwin's theory of evolution and the work of the Austrian physicist Ludwig Boltzmann on a statistical interpretation of thermodynamic processes. Physicists endeavored to describe many-particle systems like gases or fluids with statistical methods. Those statistics however, were seen as an auxiliary in situations where it was impossible for practical reasons to collect detailed information about all the properties of the system (for example all the positions and velocities of the particles of gas at a given moment in time).

The 19th century saw the development of statistics, first in the social sciences, then in physics, for example in the theory of gases, where randomness and determinacy can both be seen in the movement of molecules. On the one hand, individual molecules seem to move in an entirely random manner. On the other hand, very large numbers of the molecules which make up a gas are seen to behave in a way that obeys precise dynamical laws. How to explain this contradiction? If the movement of its constituent molecules is random and therefore cannot be predicted, surely the behavior of a gas ought to be similarly unpredictable? Yet this is far from the case.

The answer to the problem is supplied by the law of the transformation of quantity into quality. Out of the apparently random movement of a large number of molecules, there arises a regularity and a pattern which can be expressed as a scientific law. Out of chaos arises order. This dialectical relation between freedom and necessity, between chaos and order, between randomness and determinacy was a closed book to the science of the 19th century, which regarded the laws governing random phenomena (statistics) to be entirely separate and apart from the precise equations of classical mechanics.

> *Any liquid or gas [writes Gleick], is a collection of individual bits, so many that they may as well be infinite. If each piece moved independently, then the fluid would have infinitely many possibilities, infinitely many "degrees of freedom" in the jargon, and the equations describing the motion would have to deal with infinitely many variables. But each particle does not move independently — its motion depends very much on the motion of its neighbors — and in a smooth flow, the degrees of freedom can be few.* [15]

Classical mechanics worked very well for a long time, making important technological advances possible. Even down to the present time, it has a vast amount of applications. However, eventually it was found that certain areas could not adequately by dealt with by these methods. They had reached their limit. The neatly ordered, logical world of classical mechanics describes part of nature. But only part. In nature we see order, but also disorder. Alongside organization and stability there are equally powerful forces tending in the opposite direction. Here we have to resort to dialectics, to determine the relation between necessity and chance, to show at what point the accumulation of tiny, apparently insignificant changes of quantity became transformed into sudden qualitative leaps.

Bohm proposed a radical re-thinking of quantum mechanics, and a new way of looking at the relation between whole and parts.

> In these studies...it became clear that even the one-body system has a basically non-mechanical feature, in the sense that it and its environment have to be understood as an undivided whole, in which the usual classical analysis into system plus environment, considered as separately external, is no longer applicable. [The relationship of the parts] depends crucially on the state of the whole, in a way that is not expressible in terms of properties of the parts alone. Indeed, the parts are organized in ways that flow out of the whole. [16]

The dialectical law of transformation of quantity into quality expresses the idea that matter behaves differently at different levels. Thus, we have the molecular level, the laws of which are studied mainly in chemistry but partly in physics; we have the level of living matter, studied mainly in biology; the subatomic level, studied in quantum mechanics; and also another level still deeper than that of elementary particles, which is presently being explored in particle physics. Each of these levels has many subdivisions.

It has been shown that the laws governing the behavior of matter at each level are not the same. This was already shown in the 19th century by the kinetic theory of gases. If we take a box of gas containing billions of molecules, moving in irregular paths and in constant collision with other molecules, it is clearly impossible to determine the precise motions of each individual molecule. In the first place, it is ruled out on purely mathematical grounds. However, even if it were possible to solve the mathematical problems involved, it would be impossible in practice to measure the initial position and velocity of each molecule which would be needed to make precise predictions concerning it. Even a slight change in the initial angle of motion of any molecule would alter its direction, in turn leading to a still bigger change in the next collision, and so on, leading to huge errors in any prediction concerning the movement of an individual molecule.

If we try to apply the same kind of reasoning to the behavior of gases at the macroscopic ("normal") level, one would assume that it is also impossible to predict their behavior. But this is not the case, the behavior of gases at a large-scale level can be perfectly predicted. As Bohm points out:

> It is clear that one is justified in speaking of a macroscopic level possessing a set of relatively autonomous qualities and satisfying a set of relatively autonomous relations which effectively constitute a set of macroscopic casual laws. For example, if we consider a mass of water, we know by direct large-scale experience that it acts in its own characteristic way as a liquid. By this we mean that it shows all the macroscopic qualities that we associate with liquidity. For example, it flows, it 'wets' things, it tends to maintain a certain volume, etc. In its motion it satisfies a set of basic hydrodynamic equations which are expressed in terms of the large-scale properties alone, such as pressure, temperature, local density, local stream velocity, etc. Thus, if one wishes to understand the properties of the mass of water, one does not treat it as an aggregate of molecules, but rather as an entity existing at the macroscopic level, following laws appropriate to that level.

This is not to say that the molecular constitution has nothing to do with the behavior of water. On the contrary. The relation between the molecules determines, for example, whether it manifests itself as a liquid, a solid or vapor. But, as Bohm points out, there is a relative autonomy, which means that matter behaves differently at different levels; there exists "a certain stability of the characteristic modes of macroscopic behavior, which tend to maintain themselves not only more or less independently of what the individual molecules are doing, but also of the various disturbances to which the system may be subjected from outside." [17]

Is Prediction Possible?

When we toss a coin in the air, the chance that it will land "heads or tails" may be put at 50:50. That is a truly random phenomenon, which cannot be predicted. (Incidentally, when spinning, the coin is neither "heads" nor "tails"; dialectics — and the new physics — would say that it is both heads and tails.) As there are only two possible results, chance predominates. But matters change radically when very large numbers are involved. The owners of casinos, which are supposedly based on a game of "chance" know that, in the long run, zero or double zero will come up as frequently as any other number, and therefore they can make a handsome and predictable profit. The same is true of insurance companies which make a lot of money out of precise probabilities, which, in the last analysis, turn out to be practical certainties, even though the precise fate of individual clients cannot be predicted.

What are known as *"mass random events"* can be applied to a very wide field in physical, chemical, biological and social phenomena, from the sex of babies to the frequency of defects on a factory production line. The laws of probability have a very long history and have been used in the past in different spheres: the theory of errors (Gauss), the theory of accuracy in shooting (Poisson, Laplace), and above all, in statistics. For example, the *"law of great numbers"* establishes the general principle that the combined effect of a large number of accidental factors produces, for a very large class of such factors, results that are almost independent of chance. This idea was expressed as early as 1713 by Bernoulli, whose theory was generalized by Poisson in 1837, and given its final form by Chebyshev in 1867. All Heisenberg did was to apply the already known mathematics of mass-scale random events to the movements of subatomic particles, where, predictably, the element of randomness was quickly overcome.

> *Quantum mechanics having discovered precise and wonderful laws governing the probabilities, it is with numbers such as these that science overcomes its handicap of basic indeterminacy. It is by these means that science boldly predicts. Though now humbly confessing itself powerless to foretell the exact behavior of individual electrons or photons or other fundamental entities, it can yet tell you with enormous confidence how such great multitudes of them must behave precisely.* [18]

Out of apparent randomness, a pattern emerges. It is the search for such patterns, that is, for underlying laws, which forms the basis of the whole history of science. Of course, if we were to accept that everything is just random, that there is no causality, and that, anyway, we cannot know anything because there are objective limitations to our knowledge, then all will have been a complete waste of time. Fortunately, the whole history of science demonstrates that such fears are without the slightest basis. In the great majority of scientific observations, the degree of indeterminacy is so small that, for practical purposes, it may be ignored. At the level of everyday objects, the uncertainty principle proves to be absolutely useless. Thus, all the attempts to draw general philosophical conclusions from it, and apply it to knowledge and science in general, is simply a dishonest trick. Even at the subatomic level, it does not at all mean that we cannot make definite predictions. On the contrary, quantum mechanics makes very exact predictions. It is impossible to achieve a high level of certainty about the coordinates of individual particles, which may thus be said to be random. Yet, at the end of the day, out of randomness arises order and uniformity.

Accident, chance, contingencies, etc. are phenomena which cannot be defined solely in terms of the known properties of the objects under consideration. However, this does not mean that they cannot be understood. Let us consider a typical example of a chance event — a car accident. An individual accident is determined by an infinite number of chance events: if the driver had left home one minute later, if he had not turned his head for a split second, if he had been traveling ten miles an hour slower, if the old lady had not stepped into the road, etc., etc. We have all heard this kind of thing many times. The number of causes here is literally infinite. Precisely for that reason, the event is entirely unpredictable. It is accidental, and not necessary, because it might or might not have occurred. Such events, contrary to the theory of Laplace, are determined by so many independent factors that they cannot be determined at all.

However, when we consider a very large number of such accidents, the picture changes radically. There are regular trends, which can be precisely calculated and predicted by what are called statistical laws. We cannot predict an individual accident, but we can predict with great accuracy the number of accidents that will occur in a city over a period of time. Not only that, but we can introduce laws and regulations which have a definite impact on the number of accidents. Thus, there are laws which govern chance, which are just as necessary as the laws of causality themselves.

The real relationship between causality and chance was worked out by Hegel, who explained that necessity expresses itself through chance. A good example of this is the origin of life itself. The Russian scientist Oparin explains how in the complex conditions of the early period of the earth's history, the random movements of molecules would tend to form ever more complex molecules with all sorts of chance combinations. At a certain point, this huge number of accidental combinations gave rise to a qualitative leap, the emergence of living matter. At this point, the process would no longer be a matter of pure chance. Living matter would begin to evolve in accordance with certain laws, reflecting changing conditions. This relationship between the necessity and accident in science has been explored by David Bohm:

We see, then, the important role of chance. For given enough time, it makes possible, and indeed even inevitable, all kinds of combinations of things. One of those combinations which set in motion irreversible processes or lines of development that remove the system from the influence of the chance fluctuations is then eventually certain to occur. Thus, one of the effects of chance is to help "stir things up" in such a way as to permit the initiation of qualitatively new lines of development.

Polemicizing against the subjective idealist interpretation of quantum mechanics, Bohm shows conclusively the dialectal relationship between causality and chance. The existence of causality has been demonstrated by the whole history of human thought. This is not a question of philosophical speculation, but of practice and the never-ending process of human cognition:

The causal laws in a specific problem cannot be known a priori; they must be found in nature. However, in response to scientific experience over many generations along with the general background of common human experience over countless centuries, there have evolved fairly well-defined methods for finding the causal laws. The first thing that suggests causal laws is, of course, the existence of a regular relationship that holds within a wide range of variations of conditions. When we find such regularities, we do not suppose that they have arisen in an arbitrary, capricious, or coincidental fashion, but,...we assume, at least provisionally, that they are the result of necessary causal relationships. And even with regard to the irregularities, which always exist along with the regularities, one is led on the basis of general scientific experience to expect that phenomena that may seem completely irregular to us in the context of a particular stage of development of our understanding will later be seen to contain more subtle types of regularity, which will in turn suggest the existence of deeper causal relationships. [19]

Hegel on Necessity and Accident

In analyzing the nature of being in all its different manifestations, Hegel deals with the relation between potential and actual, and also between necessity and accident ("contingency"). In relation to this question, it is important to clarify one of Hegel's most famous (or notorious) sayings: *"What is rational is actual, and what is actual is rational."* [20] At first sight, this statement seems mystifying, and also reactionary, since it seems to imply that all that is exists is rational, and therefore justified. This, however, was not at all what Hegel meant, as Engels explains:

Now, according to Hegel, reality is, however, in no way an attribute predicable of any given state of affairs, social or political, in all circumstances and at all times. On the contrary. The Roman Republic was real, but so was the Roman Empire, which superseded it. In 1789 the French monarchy had become so unreal, that is to say, so robbed of all necessity, so irrational, that it had to be destroyed by the Great Revolution, of which Hegel always speaks with the greatest enthusiasm. In this case, therefore, the monarchy was the unreal and the revolution the real. And so, in the course of development, all that was previously real becomes unreal, loses its necessity, its right of existence, its rationality. And in the place of moribund reality comes a new, viable reality — peacefully if the old has enough intelligence to go to its death without a struggle; forcibly if it resists this necessity. Thus the Hegelian proposition turns into its opposite through Hegelian dialectics itself: All that is real in the sphere of human history becomes irrational in the process of time, is therefore irrational by its very destination, is tainted beforehand with irrationality; and everything which is rational in the minds of men is destined to become real, however much it may contradict existing apparent reality. In accordance with all the rules of the Hegelian method of thought, the proposition of the rationality of everything which is real resolves itself into the other proposition: All that exists deserves to perish. [21]

A given form of society is "rational" to the degree that it achieves its purpose, that is, that it develops the productive forces, raises the cultural level, and thus advances human progress. Once it fails to do this, it enters into contradiction with itself, that is, it becomes irrational and unreal, and no longer has any right to exist. Thus, even in the most apparently reactionary utterances of Hegel, there is hidden a revolutionary idea.

All that exists evidently does so of necessity. But not everything can exist. Potential existence is not yet actual existence. In *The Science of Logic*, Hegel carefully traces the process whereby something passes from a state of being merely possible to the point where possibility becomes probability, and the latter becomes inevitable ("*necessity*"). In view of the colossal confusion that has arisen in modern science around the issue of "*probability*," a study of Hegel's thorough and profound treatment of this subject is highly instructive.

Possibility and actuality denote the dialectical development of the real world and the various stages in the emergence and development of objects. A thing which exists in potential contains within itself the objective tendency of development, or at least the absence of conditions which would preclude its coming into being. However, there is a difference between abstract possibility and real potential, and the two things are frequently confused. Abstract or formal possibility merely expresses the absence of any conditions that might exclude a particular phenomenon, but it does not assume the presence of conditions which would make its appearance inevitable.

This leads to endless confusion, and is actually a kind of trick which serves to justify all kinds of absurd and arbitrary ideas. For example, it is said that if a monkey were allowed to hammer away at a typewriter for long enough, it would eventually produce one of Shakespeare's sonnets. This objective seems too modest. Why only one sonnet? Why not the collected works of Shakespeare? Indeed, why not the whole of world literature, with the theory of relativity and Beethoven's symphonies thrown in for good measure? The bare assertion that it is "statistically possible" does not take us a single step further. The complex processes of nature, society and human thought are not all susceptible to simple statistical treatment, nor will great works of literature emerge out of mere accident, no matter how long we wait for our monkey to deliver the goods.

In order for potential to become actual, a particular concatenation of circumstances is required. Moreover, this is not a simple, linear process, but a dialectical one, in which an accumulation of small quantitative changes eventually produces a qualitative leap. Real, as opposed to abstract, possibility implies the presence of all the necessary factors out of which the potential will lose its character of provisionality, and become actual. And, as Hegel explains, it will remain actual only for as long as these conditions exist, and no longer. This is true whether we are referring to the life of an individual, a given socioeconomic form, a scientific theory, or any natural phenomenon. The point at which a change becomes inevitable can be determined by the method invented by Hegel and known as the "nodal line of measurement." If we regard any process as a line, it will be seen that there are specific points (*"nodal points"*) on the line of development, where the process experiences a sudden acceleration, or qualitative leap.

It is easy to identify cause and effect in isolated cases, as when one hits a ball with a bat. But in a wider sense, the notion of causality becomes far more complicated. Individual causes and effects become lost in a vast ocean of interaction, where cause becomes transformed into effect and vice versa. Just try tracing back even the simplest event to its *"ultimate causes"* and you will see that eternity will not be long enough to do it. There will always be some new cause, and that in turn will have to be explained, and so on ad infinitum. This paradox has entered the popular consciousness in such sayings as this one:

> *For the want of a nail, a shoe was lost;*
> *For the want of a shoe, a horse was lost;*
> *For the want of a horse, a rider was lost;*
> *For the want of a rider, a battle was lost;*
> *For the want of a battle, a kingdom was lost;*
> *...And all for the want of a nail.*

The impossibility of establishing a *"final cause"* has led some people to abandon the idea of cause altogether. Everything is considered to be random and accidental. In the 20th century this position has been adopted, at least in theory, by a large number of scientists on the basis of an incorrect interpretation of the results of quantum physics, particularly the philosophical positions of Heisenberg. Hegel answered these arguments in advance, when he explained the dialectical relation between accident and necessity.

Hegel explains that there is no such thing as causality in the sense of an isolated cause and effect. Every effect has a counter-effect, and every action has a counter-action. The idea of an isolated cause and effect is an abstraction taken from classical Newtonian physics, which Hegel was highly critical of, although it enjoyed tremendous prestige at that time. Here again, Hegel was in advance of his time. Instead of the action-reaction of mechanics, he advanced the notion of Reciprocity, of universal interaction. Everything influences everything else, and is in turn, influenced and determined by everything. Hegel thus re-introduced the concept of accident which had been rigorously banned from science by the mechanist philosophy of Newton and Laplace.

At first sight, we seem to be lost in a vast number of accidents. But this confusion is only apparent. The accidental phenomena which constantly flash in and out of existence, like the waves on the face of an ocean, express a deeper process, which is not accidental but necessary. At a decisive point, this necessity reveals itself through accident. This idea of the dialectical unity of necessity and accident may seem strange, but it is strikingly confirmed by a whole series of observations from the most varied fields of science and society. The mechanism of natural selection in the theory of evolution is the best-known example. But there are many others. In the last few years, there have been many discoveries in the field of chaos and complexity theory which precisely detail how *"order arises out of chaos,"* which is exactly what Hegel worked out one and a half centuries earlier.

We must remember that Hegel was writing at the beginning of the last century, when science was completely dominated by classical mechanical physics, and half a century before Darwin developed the idea of natural selection through the medium of random mutations. He had no scientific evidence to back up his theory that necessity expresses itself through accident. But that is the central idea behind the most recent innovative thinking in science.

This profound law is equally fundamental to an understanding of history. As Marx wrote to Kugelmann in 1871:

> *World history would indeed be easy to make if the struggle were to be taken up only on condition of infallibly favorable chances. It would on the other hand be of a very mystical nature, if "accidents" played no part. These accidents naturally form part of the general course of development and are compensated by other accidents. But acceleration and delay are very much dependent upon such "accidents," including the "accident" of the character of the people who head the movement.* [22]

Engels made the same point a few years later in relation to the role of *"great men"* in history:

> *Men make their history themselves, but not as yet with a collective will according to a collective plan or even in a definite delimited given society. Their aspirations clash, and for that very reason all such societies are governed by necessity, the complement and form of appearance of which is accident. The necessity which here asserts itself athwart all accident is again ultimately economic necessity. This is where the so-called great men come in for treatment. That such and such a man and precisely that man arises at a particular time in a particular country is, of course, pure chance.*

But cut him out and there will be a demand for a substitute, and this substitute will be found, good or bad, but in the long run he will be found. [23]

Determinism and Chaos

Chaos theory deals with processes in nature that are apparently chaotic or random. A dictionary definition of chaos might suggest disorder, confusion, randomness, or chance: haphazard movement without aim, purpose or principle. But the intervention of pure "chance" into material processes invites the entry of non-physical, that is, metaphysical factors: whim, spirit or divine intervention. Because it deals with "*chance*" events, therefore, the new science of chaos has profound philosophical implications.

Natural processes which were previously considered to be random and chaotic have now proved to be lawful in a scientific sense, implying a basis in deterministic causes. Moreover, this discovery has such a widespread, not to say universal application, that it has engendered a whole new science — the study of chaos. It has created a new outlook and methodology, some would say a revolution, applicable to all established sciences. When a block of metal becomes magnetized, it goes into an "ordered state," in which all of its particles point the same way. It can be oriented one way or the other. Theoretically, it is "*free*" to orient in any direction. In practice, every little piece of metal makes the same "*decision.*"

A chaos scientist has worked out the basic mathematical rules that describe the "*fractal geometry*" of a leaf of the black spleenwort fern. He has fed the information into his computer which also has a random number generator. It is programmed to build up a picture using dots put at random on the screen. As the experiment progresses, it is impossible to anticipate where each dot will appear. But unerringly, the image of the fern leaf is built up. The superficial similarity between these two experiments is obvious. But it suggests a deeper parallel. Just as the computer was basing its apparently random selection of dots (and to the observer "outside" the computer, for all practical purposes it was random) on well-defined mathematical rules, so also it would suggest that the behavior of photons (and by implication all quantum events) are subject to underlying mathematical rules which, however, are well beyond human understanding at the present time.

The Marxist view holds that the entire universe is based upon material forces and processes. Human consciousness is in the final analysis only a reflection of the real world that exists outside it, a reflection based on the physical interaction between the human body and the material world. In the material world there is no discontinuity, no interruption in the physical interconnection of events and processes. There is no room, in other words, for the intervention of metaphysical or spiritual forces. Materialist dialectics, Engels said, is the "*science of universal interconnection.*" Moreover, the interconnectedness of the physical world is based upon the principle of causality, in the sense that processes and events, are determined by their conditions and the lawfulness of their interconnections:

> *The first thing that strikes us in considering matter in motion is the interconnection of the individual motions of separate bodies, their being determined by one another. But not only do we find that a particular motion is followed by another, we find also that we can evoke a particular motion by setting up the conditions in which it takes place in nature, that we can even produce motions which do not occur at all in nature (industry), at least not in this way, and that we can give these motions a predetermined direction and extent. In this way, by the activity of human beings, the idea of causality becomes established, the idea that one motion is the cause of another.* [24]

The complexity of the world may disguise the processes of cause and effect and make the one indistinguishable from the other, but that does not alter the underlying logic. As Engels explained, "*cause and effect are conceptions which only hold good their application to individual cases; but as soon as we consider the individual cases in their general interconnection with the universe as a whole, they run into each other, and they become confounded when we contemplate that universal action and reaction in which causes and effects are eternally changing places, so that what is effect here and now will be cause there and then, and vice versa.*" [25]

Chaos theory undoubtedly represents a big advance, but here also there are some questionable formulations. The celebrated butterfly effect, according to which a butterfly flaps its wings in Tokyo, and causes a storm the following week in Chicago is no doubt a sensational example, intended to provoke controversy. However, it is incorrect in this form. Qualitative changes can only occur as the result of an accumulation of quantitative changes. A small accidental change (a butterfly flapping its wings) could only produce a dramatic result if all the conditions for a storm were already in existence. In this case, necessity could express itself through an accident. But only in this case.

The dialectical relationship between necessity and chance can be seen in the process of natural selection. The number of random mutations within the organism is infinitely large. However, in a particular environment, one of these mutations is found to be useful to the organism and retained, while all the others, perish. Necessity once again manifests itself through the agency of chance. In a sense, the appearance of life on earth can be seen as an "*accident.*" It was not preordained that the earth should be exactly at the right distance from the sun, with the right kind of gravity and atmosphere, for this to happen. But, given this concatenation of circumstances, over a period of time, out of a vast number of chemical reactions, life would inevitably arise. This applies not only to our own planet, but to a vast number of other planets where similar conditions exist, although not in our solar system. However, once life had arisen, it ceases to be a question of accident, and develops according to its own inherent laws.

Consciousness itself did not arise out of any Divine plan, but, in one sense also arose from the *"accident"* of bipedalism (upright stance), which freed the hands, and thus made it possible for early hominids to evolve as a tool-making animal. It is probable that this evolutionary quirk was the result of a climatic change in East Africa, which partly destroyed the forest habitat of our simian ancestors. This was an accident. As Engels explains in *The Part Played by Labor in the Transition of Ape to Man*, this was the basis upon which human consciousness developed. But in a broader sense, the emergence of consciousness — of matter aware of itself — cannot be regarded as an accident, but a necessary product of the evolution of matter, which proceeds from the simplest forms to more complex forms, and which, where the conditions exist, will inevitably give rise to intelligent life, and higher forms of consciousness, complex societies, and what we know as civilization.

In his *Metaphysics*, Aristotle devotes a lot of space to a discussion of the nature of necessity and accident. He gives us an example, the accidental words that lead to a quarrel. In a tense situation, for example a marriage in difficulties, even the most innocuous comment can lead to a row. But it is clear that the words spoken are not the real cause of the dispute. It is the product of an accumulation of stresses and strains, which sooner or later reaches a breaking-point. When this point is reached, the slightest thing can provoke an outburst. We can see the same phenomenon in the workplace. For years, an apparently docile workforce, fearful of unemployment, is prepared to accept all manner of impositions — wage reductions, sackings of colleagues, worsening conditions, etc. On the surface, nothing is happening. But in reality, there is a steady increase in discontent, which, at a certain point, must find an expression. One day, the workers decide that *"enough is enough."* At this precise point, even the most trivial incident can provoke a walk-out. The whole situation changes into its opposite.

There is a broad analogy between the class struggle and the conflicts between nations. In August 1914, the Crown Prince of Austro-Hungary was assassinated in Sarajevo. This was alleged to have caused the First World War. As a matter of fact, this was an historical accident which might or might not have occurred. Prior to 1914, there were several other incidents (the Morocco incident, the Agadir incident) which could equally have led to war. The real cause of World War One was the accumulation of unbearable contradictions between the main imperialist powers — Britain, France, Germany, Austro-Hungary and Russia. This reached a critical stage, where the whole explosive mixture could be ignited by a single spark in the Balkans.

Finally, we see the same phenomenon in the world of economics. At the moment when we write these lines the City of London has been shaken by the collapse of the Barings Bank. This was instantly blamed on the fraudulent activities of one of the bank's employees in Singapore. But the Barings collapse was merely the latest symptom of a far deeper malaise in the world financial system. The headlines in *The Independent* newspaper read "an accident waiting to happen." On a world scale, there are at present US $25 trillion invested in derivatives. This shows that capitalism is no longer based on production, but to a greater and greater extent upon speculative activities. The fact that Mr. Leeson lost a large amount of money in the Japanese stock markets may be connected with the accident of the Kobe earthquake. But serious economic analysts understand that this was an expression of the fundamental unsoundness of the international financial system. With or without Mr. Leeson, future collapses are inevitable. The big international corporations and financial institutions, all of whom are involved in this reckless gambling, are playing with fire. A major financial collapse is implicit in the whole situation.

It may be that there are many phenomena whose underlying processes and causative relationships are not fully understood so that they appear to be random. For all practical purposes, therefore these can only be treated statistically, like the roulette wheel to the punter. But underlying these "chance" events there are still forces and processes that determine the end results. We live in a universe governed by dialectical determinism.

Marxism and Freedom

The problem of the relation between "freedom and necessity" was known to Aristotle and endlessly discussed by the mediaeval Schoolmen. Kant uses it as one of his celebrated *"antinomies,"* where it is presented as an insoluble contradiction. In the 17th and 18th centuries it cropped up in mathematics as the theory of chance, related to gambling.

The dialectical relationship between freedom and necessity has re-surfaced in chaos theory. Doyne Farmer, an American physicist investigating complicated dynamics, comments:

> *On a philosophical level, it struck me as an operational way to define free will, in a way that allowed you to reconcile free will with determinism. The system is deterministic, but you can't say what it's going to do next. At the same time, I'd always felt that the important problems out there in the world had to do with the creation of organization, in life or intelligence. But how did you study that? What biologists were doing seemed so applied and specific; chemists certainly weren't doing it; mathematicians weren't doing it at all, and it was something that physicists just didn't do. I always felt that the spontaneous emergence of self-organization ought to be part of physics. Here was one coin with two sides. Here was order, with randomness emerging, and then one step further away was randomness with it own underlying order.* [26]

Dialectical determinism has nothing in common with the mechanical approach, still less with fatalism. In the same way that there are laws which govern inorganic and organic matter, so there are laws that govern the evolution of human society. The patterns which can be observed through history are not at all fortuitous. Marx and Engels explained that the transition from one social system to another is determined by the development of the productive forces, in the last analysis. When a given socioeconomic system is no longer able to develop the productive forces, it enters into crisis, preparing the ground for a revolutionary overturn.

This is not at all to deny the role of the individual in history. As we have already said, men and women make their own history. However, it would be foolish to imagine that human beings are *"free agents"* who can determine their future purely on the basis of their own will. They have to base themselves on conditions which have been created independent of their will — economic, social, political, religious, and cultural. In this sense, the idea of free-will is nonsense. The real attitude of Marx and Engels towards the role of the individual in history is shown by the following quotation from *The Holy Family*:

> *History does nothing, it "possesses no immense wealth," it "wages no battles." It is man, real, living man who does all that, who possesses and fights; "history" is not, as it were, a person apart, using man as a means to achieve its own aims; history is nothing but the activity of man pursuing its aims.* 27

There is no question of men and women being merely blind puppets of fate, powerless to change their own destiny. However, the real men and women living in the real world of which Marx and Engels write, do not and cannot stand above the society in which they live. Hegel once wrote that *"interests move the life of the peoples."* Consciously or otherwise, the individual actors on the historical stage ultimately reflect the interests, opinions, prejudices, morality and aspirations of a specific class or group within society. This is really self-evident from even the most superficial reading of history.

Nevertheless, the illusion of *"free-will"* is persistent. The German philosopher Leibniz remarked that a magnetic needle, if it could think, would doubtless imagine that it pointed North because it choose to do so. In the 20th century, Sigmund Freud utterly demolished the prejudice that men and women are in complete control even of their own thoughts. The phenomenon of "Freudian slips" is a perfect example of the dialectical relationship between accident and necessity. Freud gives numerous examples of mistakes in speech, *"forgetfulness,"* and other *"accidents,"* which, in many cases, undoubtedly reveal deeper psychological processes. In the words of Freud:

"Certain inadequacies of our psychic capacities...and certain performances which are unintentional prove to be well motivated when subjected to the psycho-analytic investigation, and are determined through the consciousness of unknown motives." [28]

It was a fundamental tenet of Freud's approach that none of human behavior is accidental. The small mistakes of everyday life, dreams, and the apparently inexplicable symptoms of mentally ill people are not *"accidental."* By definition, the human mind is not aware of unconscious processes. The more deeply unconscious the motivation, from the standpoint of psychoanalysis, the more obvious it is that a person will not be aware of it. Freud grasped early on the general principle that these unconscious processes reveal themselves (and therefore can be studied) in those fragments of behavior which the conscious mind dismisses as silly mistakes or accidents.

Is it possible to attain freedom? If what is meant by a "free" action is one that is not caused or determined, we must say quite frankly that such an action has never existed, and never will exist. Such imaginary "freedom" is pure metaphysics. Hegel explained that real freedom is the recognition of necessity. To the degree that men and women understand the laws that govern nature and society, they will be in a position to master these laws and turn them to their own advantage. The real material basis upon which humankind can become free has been established by the development of industry, science and technique. In a rational system of society — one in which the means of production are harmoniously planned and consciously controlled — we will really be able to speak about free human development. In the words of Engels, this is "mankind's leap from the realm of necessity to the realm of freedom."

7.　RELATIVITY THEORY

What is Time?

Few ideas have penetrated the human consciousness as profoundly as that of time. The idea of time and space has occupied human thought for thousands of years. These things at first sight seem simple and easy to grasp, because they are close to everyday experience. Everything exists in time and space, so they appear as familiar conceptions. However, what is familiar is not necessarily understood. On closer examination, time and space are not so easily grasped. In the 5th century, St. Augustine remarked: "What, then, is time? If no one asks me, I know what time is. If I wish to explain it to him who asks me, I do not know." The dictionary is not much help here. Time is defined as a "a period," and a period is defined as "time." This does not get us very far! In reality, the nature of time and space is quite a complex philosophical problem.

Men and women clearly distinguish between past and future. A sense of time is, however, not unique to humans or even animals. Organisms often have a kind of "internal clock," like plants which turn one way during the day and another at night. Time is an objective expression of the changing state of matter. This is revealed even by the way we talk about it. It is common to say that time "flows." In fact, only material fluids can flow. The very choice of metaphor shows that time is inseparable from matter. It is not only a subjective thing. It is the way we express an actual process that exists in the physical world. Time is thus just an expression of the fact that all matter exists in a state of constant change. It is the destiny and necessity of all material things to change into something other than what they are. "Everything that exists deserves to perish."

A sense of rhythm underlies everything: the heart-beat of a human, the rhythms of speech, the movement of the stars and planets, the rise and fall of the tides, the alternations of the seasons. These are deeply engraved upon the human consciousness, not as arbitrary imaginings, but as real phenomena expressing a profound truth about the universe. Here human intuition is not in error. Time is a way of expressing change of state and motion which are inseparable features of matter in all its forms. In language we have tense, future, present and past. This colossal conquest of the mind enabled humankind to free itself from the slavery of the moment, to rise above the concrete situation and be "present," not just in the here and now, but in the past and the future, at least in the mind.

Time and movement are inseparable concepts. They are essential to all life and all knowledge of the world, including every manifestation of thought and imagination. Measurement, the corner-stone of all science, would be impossible without time and space. Music and dance are based upon time. Art itself attempts to convey a sense of time and movement, which are present not just in representations of physical energy, but in design. The colors, shapes and lines of a painting guide the eye across the surface in a particular rhythm and tempo. This is what gives rise to the particular mood, idea and emotion conveyed by the work of art. Timelessness is a word that is often used to describe works of art, but really expresses the opposite of what is intended. We cannot conceive of the absence of time, since time is present in everything.

There is a difference between time and space. Space can also express change, as change of position. Matter exists and moves through space. But the number of ways that this can occur is infinite: forward, backward, up or down, to any degree. Movement in space is reversible. Movement in time is irreversible. They are two different (and indeed contradictory) ways of expressing the same fundamental property of matter — change. This is the only Absolute that exists.

Space is the "otherness" of matter, to use Hegel's terminology, whereas time is the process whereby matter (and energy, which is the same thing) constantly changes into something other than what it is. Time — "the fire in which we are all consumed" — is commonly seen as a destructive agent. But it is equally the expression of a permanent process of self-creation, whereby matter is constantly transformed into and endless number of forms. This process can be seen quite clearly in non-organic matter, above all at the subatomic level.

The notion of change, as expressed in the passing of time, deeply permeates human consciousness. It is the basis of the tragic element in literature, the feeling of sadness at the passing of life, which reaches its most beautiful expression in the sonnets of Shakespeare, like this one which vividly conveys a sense of the restless movement of time:

> *Like as the waves make toward the pebbled shore,*
> *So do our minutes hasten to their end;*
> *Each changing place with that which goes before,*
> *In sequent toil all forward do contend.*

The irreversibility of time does not only exist for living beings. Not only humans, but stars and galaxies are born and perish. Changes affects all, but not only in a negative way. Alongside death there is life, and order arises spontaneously out of chaos. The two sides of the contradiction are inseparable. Without death, life itself would be impossible. Every man and woman is not only aware of themselves, but also the negation of themselves, their limit. We come from nature and will return to nature.

Mortals understand that as finite beings their lives must end in death. As the *Book of Job* reminds us: "Man that is born of woman is of a few days, and full of trouble. He cometh forth like a flower, and is cut down; he fleeth also as a shadow, and continueth not." [29] Animals do not fear death in the same way because they have no knowledge of it. Human beings have attempted to escape their destiny by establishing a privileged communion with an imaginary supernatural existence after death. The idea of everlasting life is present in almost all religions in one form or another. It is the motive-force behind the egotistical thirsting for an imaginary immortality in a non-existent Heaven, which is supposed to provide a consolation for the "Vale of Tears" on this sinful earth. Thus, for countless centuries men and women have been taught to submit meekly to suffering and privation on earth in expectation of a life of happiness — once they are dead.

That every individual must pass away is well known. In the future, human life will be prolonged far beyond its "natural" span; nevertheless the end must come. But what is true for particular men and women is not true of the species. We live on through our children, through the memories of our friends, and through the contribution we make to the good of humanity. This is the only immortality to which we are entitled to aspire. Generations pass away, but are replaced by new generations, which develop and enrich the scope of human activity and knowledge. Humanity can conquer the earth and reach out its hands to the heavens. The real search for immortality is realized in this endless process of human development and perfection, as men and women make themselves anew on a higher basis than before. The highest goal we can set ourselves is thus not to long for an imaginary paradise in the beyond, but to fight to attain the real social conditions for the building of a paradise in this world.

From our earliest experiences, we come to an understanding of the importance of time. So it is surprising that some have thought time to be an illusion, a mere invention of the mind. This idea has persisted down to the present In fact, the idea that time and change are mere illusions is not new. It is present in ancient religions like Buddhism, and also in idealist philosophies like that of Pythagoras, Plato and Plotinus. The aspirations of Buddhism was to reach Nirvana, a state where time ceased to exist. It was Heraclitus, the father of dialectics, who understood correctly the nature of time and change, when he wrote that "everything is and is not because everything is in flux" and "we step and do not step in the same stream, we are and are not."

The idea of change as cyclical is the product of an agricultural society utterly dependent upon the change of seasons. The static way of life rooted in the mode of production of former societies found its expression in static philosophies. The Catholic Church could not stomach the cosmology of Copernicus and Galileo because it challenged the existing view of the world and society. Only in capitalist society has the development of industry disrupted the old, slow rhythms of peasant life. Not only is the difference between the seasons abolished in production, but even the difference between night and day, as machines run for 24 hours a day, seven days a week, fifty two weeks a year, under the glare of artificial lights. Capitalism has revolutionized the means of production, and with it the minds of men and women. However, the progress of the latter has proved to be far slower than the former. The conservatism of the mind is revealed in the constant attempt to cling to outworn ideas, old certainties whose time has long past, and, ultimately, the age-old hope for a life after death.

The idea that universe must have a beginning and an end has been revived in recent decades by the cosmological theories of the big bang. This inevitably involves a supernatural being who creates the world according to some unfathomable plan from nothing, and keeps it going for as long as He considers it necessary. The old religious cosmology of Moses, Isaiah, Tertullian and Plato's *Timaeus*, incredibly resurfaces in the writings of some modern cosmologists and theoretical physicists. There is nothing new in this. Every social system which enters into a phase of irreversible decline always presents its own demise as the end of the world, or, better still, the universe. Yet the universe still carries on, indifferent to the destiny of this or that temporary social formation on earth. Humankind continues to live, to fight and, despite all reverses, to develop and progress. So that every period sets out on a higher level than before. And there is, in principle, no limit to this process.

Time and Philosophy

The Ancient Greeks actually had a far deeper insight into the meaning of time, space and motion than the moderns. Not only Heraclitus, the greatest dialectician of Antiquity, but also the Eleatic philosophers (Parmenides, Zeno) arrived at a very scientific conception of these phenomena. The Greek atomists already put forward the picture of a universe which required no Creator, no beginning and no end. Space and matter are generally seen as opposites, as conveyed by the idea of "full" and "empty." In practice, however, the one cannot exist without the other. They presuppose each other, determine, limit and define each other. The unity of space and matter is the most fundamental unity of opposites of all. This was already understood by the Greek atomists who visualized the universe as being composed of only two things — the "atoms" and the "void." In essence, this view of the universe is correct.

Relativism has been observed many times in the history of philosophy. The sophists held that "man is the measure of all things." They were relativists par excellence. Denying the possibility of absolute truth, they inclined towards extreme subjectivism. The sophists nowadays have a bad name, but in fact they represented a step forward in the history of philosophy. While there were many charlatans in their ranks, they also had a number of talented dialecticians like Protagoras. The dialectic of sophism was based on the correct idea that truth is many sided. A thing can be shown to have many properties. It is necessary to have the ability to see a given phenomenon from different sides. For the undialectical thinker, the world is a very simple place, made up of things existing separately, one after the other. Every "thing" enjoy a solid existence in time and space. It is before me "here" and "now." However, closer observation reveals these simple and familiar words to be one-sided abstractions.

Aristotle as in so many other fields dealt with space, time and motion with great rigor and profundity. He wrote that only two things are imperishable: time and change, which he rightly considers identical:

> It is impossible, however, that motion should be generable or perishable; it must always have existed. Nor can time come into being or cease to be; for there cannot be a "before" or "after" where there is no time. Movement, then, is also continuous in the sense in which time is, for time is either the same thing as motion or an attribute

of it; so then motion must be continuous as time is, and if so it must be local and circular. [Elsewhere he says that] Movement can neither come into being nor cease to be: nor can time come into being, or cease to be. [30]

How much wiser were the great thinkers of the Ancient World than those who now write about "the beginning of time," and without even smiling!

The German idealist philosopher Emmanuel Kant was the man who, after Aristotle, investigated the question of the nature of time and space most fully, although his solutions were ultimately unsatisfactory. Every material thing is an assemblage of many properties. If we take away all these concrete properties, we are left with only two abstractions: time and space. The idea of time and space as really existing metaphysical entities was given a philosophical basis by Kant, who claimed that space and time were "phenomenally real," but could not be known "in themselves."

Time and space are properties of matter, and cannot be conceived separately from matter. In his book *The Critique of Pure Reason*, Kant claimed that time and space were not objective concepts drawn from observation of the real world, but were somehow inborn. In point of fact, all the concepts of geometry are derived from observations of material objects. One of the achievements of Einstein's general theory of relativity was precisely to develop geometry as an empirical science, the axioms of which are inferred from actual measurements, and which differ from the axioms of classical Euclidean geometry, which were (incorrectly) supposed to have been the products of pure reason, deduced from logic alone.

Kant attempted to justify his claims in the famous section in his *Critique of Pure Reason* known as the *Antinomies*. which deal with the contradictory phenomena of the natural world, including space and time. The first four of Kant's (cosmological) antinomies deal with this question. Kant had the merit of posing the existence of such contradictions, but his explanation was at best incomplete. It fell to the great dialectician Hegel to resolve the contradiction in *The Science of Logic*.

Throughout the 18th century, science was dominated by the theories of classical mechanics, and one man set his stamp on the whole epoch. The poet Alexander Pope sums up the adulatory attitude of contemporaries to Newton in his verse:

Nature and Nature's laws lay hid in night:
God said "Let Newton be!" and all was light.

Newton envisaged time as flowing in a straight line everywhere. Even if there was no matter, there would be a fixed frame of space and time would still flow "through" it. Newton's absolute spatial frame was supposed to be filled with a hypothetical "ether" through which light waves flowed. Newton thought that time was like a gigantic "container" inside which everything exists and changes. In this idea, time is conceived as having an existence separate and apart from the natural universe. Time would exist, even if the universe did not. This is characteristic of the mechanical (and idealist) method in which time, space, matter and motion are regarded as absolutely separate. In reality, it is impossible to separate them.

Newtonian physics was conditioned by mechanics, which in the 18th century was the most advanced of the sciences. It was also convenient for the new ruling class because it presented an essentially static, timeless, unchanging view of the universe, in which all contradiction were smoothed out — no sudden leaps, no revolutions, but a perfect harmony, in which everything sooner or later returned to equilibrium, just as the British parliament had reached a satisfactory equilibrium with the Monarchy under William of Orange. The 20th century has pitilessly destroyed this view of the world. One after the other, the old rigid, static mechanism has been displaced. The new science has been characterized by restless change, fantastic speed, contradictions and paradoxes at all levels.

Newton distinguished between absolute time and "relative, apparent and common time," as it appears in earthly clocks. He advanced the notion of absolute time, an ideal time scale which simplified the laws of mechanics. These abstractions of time and space proved to be powerful ideas which have greatly advanced our understanding of the universe. They were held to be absolute for a long time. However, upon closer examination, the "absolute truths" of classical Newtonian mechanics proved to be — relative. They were true only within certain limits.

Newton and Hegel

The mechanistic theories which dominated science for two centuries after Newton, were first seriously challenged in the field of biology by the revolutionary discoveries of Charles Darwin. Darwin's theory of evolution showed that life could originate and develop without the need for Divine intervention, on the basis of the laws of nature. At the end of the 19th century, the idea of the "arrow of time" was put forward by Ludwig Boltzmann in the second law of thermodynamics. This striking image no longer presents time as a never-ending cycle, but as an arrow moving in a single direction. These theories assume that time is real and that the universe is in a continual process of change, as old Heraclitus had foreseen.

Almost half a century before Darwin's epoch making work, Hegel had anticipated not only him, but many other discoveries of modern science. Boldly challenging the assumption of the prevailing Newtonian mechanics, Hegel advanced a dynamic view of the world, based on processes and change through contradiction. The brilliant anticipations of Heraclitus were transformed by Hegel into a completely elaborated system of dialectical thought. There is no doubt that, had Hegel been taken more seriously, the process of science would have advanced far more rapidly than it did.

The greatness of Einstein was to get beyond these abstractions and reveal their relative character. The relative aspect of time was, however, not new. It was thoroughly analyzed by Hegel. In his early work The Phenomenology of Mind, he explains the relative content of words like "here" and "now." These ideas which seem quite simple and straightforward turn out to be very complex and contradictory. "To the question, What is the Now? we reply, for example, the Now is night-time. To test the truth of this certainty of sense, a simple experiment is all we need: write that truth down. A truth cannot lose anything by being written down, and just as little by our preserving and keeping it. If we look again at the truth we have written down, look at it now, at his noon-time, we shall have to say it has turned stale and become out of date." [31]

It is a very simple matter to dismiss Hegel (or Engels) because their writings on science were necessarily limited by the actual state of science of the day. What is remarkable, however, is how advanced Hegel's views on science actually were. In their book *Order out of Chaos*, Prigogine and Stengers point out that Hegel rejected the mechanistic method of classical Newtonian physics, at a time when Newton's ideas were universally sacrosanct:

> The Hegelian philosophy of nature systematically incorporates all that is denied by Newtonian science. In particular, it rests on the qualitative difference between the simple behavior described by mechanics and the behavior of more complex entities such as living beings. It denies the possibility of reducing those levels, rejecting the idea that differences are merely apparent and that nature is basically homogeneous and simple. It affirms the existence of a hierarchy, each level of which presupposes the preceding ones. [32]

Hegel wrote scornfully about the allegedly absolute truths of Newtonian mechanics. He was the first one to subject the mechanistic approach of the 18th century to a thorough criticism, although the limitations of the science of his day did not allow him to put forward a worked-out alternative. For Hegel, every finite thing was mediated, that is, relative to something else. Moreover, this relationship was not merely a formal juxtaposition, but a living process: everything was limited, conditioned and determined by everything else. Thus, cause and effect only hold good in relation to isolated relations (such as we find in classical mechanics), but not if we regard things as processes, in which everything is the result of universal interrelations and interactions.

Time is the form of existence of matter. Mathematics and formal logic cannot really deal with time, but treat it merely as a quantitative relation. Now there is no doubt about the importance of quantitative relations for understanding reality, since every finite thing can be approached from a quantitative point of view. Without a grasp of quantitative relationships, science would be impossible. But in and of themselves, they cannot adequately express the complexity of life and movement, the restless process of change in which gradual, smooth developments suddenly give rise to chaotic transformations.

Purely quantitative relations, to use Hegel's terminology, present the real processes of nature "only in an arrested paralyzed form." [33] The universe is an infinite, self-moving whole, which is self-establishing and contains life within itself. Movement is a contradictory phenomenon, containing both positive and negative. This is one of the fundamental propositions of dialectics, which are closer to the real nature of things than the axioms of classical mathematics.

Only in classical geometry is it possible to conceive of completely empty space. It is yet another mathematical abstraction, which plays an important role, but only approximately represents reality. Geometry essentially compares different spatial magnitudes. Contrary to what Kant believed, the abstractions of mathematics are not "a priori" and inborn, but derived from observations of the material world. Hegel shows that the Greeks had already understood the limitedness of purely quantitative descriptions of nature, and comments:

> How much further had they progressed in thought than those who in our day, when some put in the place of determinations of thought number and determinations of numbers (like powers), next the infinitely great and the infinitely small, one divided by infinity, and other such determinations, which often are a perverted mathematical formalism, take the return to this impotent childishness for something praiseworthy and even for something thorough and profound. [34]

These lines are even more appropriate today than when they were written. It really is incredible when certain cosmologists and mathematicians make the most preposterous claims about the nature of the universe without the slightest attempt to prove them on the basis of observed facts, and then appeal to the alleged beauty and simplicity of their equations as the final authority. The cult of mathematics is greater today than at any time since Pythagoras who thought that "all things are Number." And, as with Pythagoras, there are similarly mystical overtones. Mathematics leaves aside all qualitative determinations except number. It ignores the real content, and applies its rules externally to things. None of these abstractions have real existence. Only the material world exists. This fact is all too frequently overlooked with disastrous results.

Relativity

Albert Einstein was undoubtedly one of the great geniuses of our time. Between his twenty first and thirty eighth birthdays he completed a revolution in science, with profound repercussions at many levels. The two great breakthroughs were the Special Theory of Relativity (1905) and the General Theory of Relativity (1915). Special relativity deals with high speeds, general relativity with gravity.

Despite their extremely abstract character, Einstein's theories were ultimately derived from experiments, and were successfully given practical applications, which confirmed their correctness time and again. Einstein set out from the famous Michelson-Morley experiment, "the greatest negative experiment of the history of science" (Bernal), which exposed an inner contra-diction in 19th century physics. This experiment attempted to generalize the electromagnetic theory of light by demonstrating that the apparent velocity of light was dependent upon the rate at which the observer traveled through the supposedly fixed "ether." In the end, no difference was found in the velocity of light, in whatever direction the observer was traveling.

J. J. Thomson later showed that the velocity of electrons in high electrical fields was slower than predicted by the classical Newtonian physics. These contradictions in 19th century physics were resolved by the special theory of relativity. The old physics was unable to explain the phenomenon of radioac-tivity. Einstein explained this as the release of a tiny part of the enormous amount of energy trapped in "inert" matter.

In 1905, Einstein developed his special theory of relativity in his spare time, while working as a clerk in a Swiss patent office. Setting out from the discoveries of the new quantum mechanics, he showed that light travels through space in a quantum form (as bundles of energy). This was clearly in contradiction to the previously accepted theory of light as a wave. In effect, Einstein revived the old corpuscular theory of light, but in an entirely different way. Here light was shown as a new kind of particle, with a contradictory character, simultaneously displaying the properties of a particle and a wave. This startling theory made possible the retention of all the great discoveries of 19th century optics, including spectroscopes, as well as Maxwell's equation. But it killed stone-dead the old idea that light requires a special vehicle, the "ether," to travel through space.

Special relativity starts from the assumption that the speed of light in a vacuum will always be measured at the same constant value, irrespective of the speed of the light source relative to the observer. From this it is deduced that the speed of light represents the limiting speed for anything in the universe. In addition, special relativity states that energy and mass are in reality equivalents. This is a striking confirmation of the fundamental philosophical postulate of dialectical materialism — the inseparable character of matter and energy the idea that motion ("energy") is the mode of existence of matter.

Einstein's discovery of the law of equivalence of mass and energy is expressed in his famous equation $E = mc^2$, which expresses the colossal energies locked up in the atom. This is the source of all the concentrated energy in the universe. The symbol e represents energy (in ergs), m stands for mass (in grams) and c is the speed of light (in centimeters per second). The actual value of c^2 is 900 billion billion. That is to say, the conversion of one gram of energy locked up in matter will produce a staggering 900 billion billion ergs. To give a concrete example of what this means, the energy contained in a single gram of matter is equivalent to the energy produced by burning 2,000 tons of petrol.

Mass and energy are not just "interchangeable," as dollars are interchangeable with euros, they are one and the same substance, which Einstein characterized as "mass-energy." This idea goes far deeper and is more precise than the old mechanical concept whereby, for example, friction is transformed into heat. Here, matter is just a particular form of "frozen" energy, while every other form of energy (including light), has mass associated with it. For this reason, it is quite wrong to say that matter "disappears" when it is changed into energy.

Einstein's law displaced the old law of the conservation of mass, worked out by Lavoisier, which says that matter, understood as mass, can neither be created nor destroyed. In fact, every chemical reaction that releases energy converts a small amount of mass into energy. This could not be measured in the kind of chemical reaction known to the 19th century, such as the burning of coal. But nuclear reaction releases sufficient energy to reveal a measurable loss of mass. All matter, even when at "rest," contains staggering amounts of energy. However, as this cannot be observed, it was not understood until Einstein explained it.

Far from overthrowing materialism, Einstein's theory establishes it on a firmer basis. In place of the old mechanical law of the "conservation of mass," we have the far more scientific and more general laws of the conservation of mass-energy, which expresses the first law of thermodynamics in an universal and unassailable form. The mass does not "disappear" at all, but is converted into energy. The total amount of mass-energy remains the same. Not a single particle of matter can be created or destroyed. The second idea is the special limiting character of the speed of light: the assertion that no particle can travel faster than the speed of light, since as it approaches this critical velocity, its mass approaches infinity, so that it becomes harder and harder to go faster. These ideas seem abstract and difficult to grasp. They challenge the assumptions of "sound common sense."

The relationship between "common sense" and science was summed up by the Soviet scientist Professor L. D. Landau in the following lines: "So-called common sense represents nothing but a simple generalization of the notions and habits that have grown up in our daily life. It is a definite level of understanding reflecting a particular level of experiment." And he adds, "Science is not afraid of clashes with so-called common sense. It is only afraid of disagreement between existing ideas and new experimental facts and if such disagreement occurs science relentlessly smashes the ideas it has previously built up and raises our knowledge to a higher level."[35] How can a moving object increase its mass? Such a notion contradicts our everyday experience. A spinning top does not visibly gain in mass while revolving. In point of fact, it does, but the increase is so infinitesimal that it may be discounted for all practical purposes. The effects of special relativity cannot be observed on the level of everyday phenomena. However, under extreme conditions, for example, at very high speeds approaching the speed of light, relativistic effects begin to come into play.

Einstein predicted that the mass of a moving object would increase at very high speeds. This law can be ignored when dealing with normal speeds. Nevertheless, subatomic particles move at speeds of nearly 10,000 miles per second or more, and at such speeds as these relativistic effects appear. The discoveries of quantum mechanics demonstrated the correctness of the special theory of relativity, not only qualitatively, but quantitatively. An electron gains in mass as it moves at 9/10th the speed of light; moreover, the gain in mass is 3 1/6th times, precisely as Einstein's theory predicted. Since then, special relativity has been tested many times, and so far it has always given correct results. Electrons emerge from a powerful particle accelerator about 40,000 times heavier than when they started, the extra mass representing energy of motion.

At far higher velocities the increase in mass becomes noticeable. And modern physics deals precisely with extremely high velocities, such as the speed of sum-atomic particles, which approach the speed of light. Here the classical laws of mechanics, which adequately describe everyday phenomena, cannot be applied. To common sense the mass of an object never changes. Therefore a spinning-top has the same weight as a still one. In this way a law was invented which states that mass is constant irrespective of speed.

Later, this law was shown to be incorrect. It was found that mass increases with velocity. Yet, since the increase only becomes appreciable near the speed of light, we take it as constant. The correct law would be: "If an object moves with a speed of less than 100 miles per second, the mass is consistent to within one part in a million." For everyday purposes, we can assume that mass is constant irrespective of speed. But for high speeds, this is false, and the higher the speed, the falser is the assertion. Like thinking based on formal logic, it is accepted as valid for practical purposes. Feynman points out:

> ...Philosophically, we are completely wrong with the approximate law. Our entire picture of the world has to be altered even though the mass changes only by a little bit. This is a very peculiar thing about the philosophy, or the ideas, behind the laws. Even a very small effect sometimes requires profound changes in our ideas. [36]

The predictions of special relativity have been shown to correspond to the observed facts. Scientists discovered by experiment that gamma-rays could produce atomic particles, transforming the energy of light into matter. They also found that the minimum energy required to create a particle depended on its rest-energy, as predicted by Einstein. In point of fact not one, but two particles were produced: a particle and its opposite, the "anti-particle." In the gamma-ray experiment, we get an electron and an anti-electron (positron). The reverse process also takes place: when a positron meets an electron, they annihilate each other, producing gamma rays. Thus, energy is transformed into matter, and matter into energy. Einstein's discovery provided the basis for a far more profound understanding of the workings of the universe. It provided an explanation of the source of the sun's energy, which had been a mystery throughout the ages. The immense storehouse of energy turned out to be — matter itself. The awesome power of the energy locked up in matter was revealed to the world in August 1945 at Hiroshima and Nagasaki. All this is contained in the deceptively simple formula $E = mc^2$.

The General Theory of Relativity

Special relativity is quite adequate when dealing with an object moving at constant speed and direction in relation to the observer. However, in practice motion is never constant. There are always forces which cause variations in the speed and direction of moving objects. Since subatomic particles move at immense speeds over short distances, they do not have time to accelerate much, and special relativity can be applied. Nevertheless, in the motion of planets and stars, special relativity proved insufficient. Here we are dealing with large accelerations caused by huge gravitational fields. It is once again a case of quantity and quality. At the subatomic level, gravitation is insignificant in comparison with other forces, and can be ignored. In the everyday world, on the contrary, all other forces except gravity can be ignored.

Einstein attempted to apply relativity to motion in general, not just to constant motion. Thus we arrive at the general theory of relativity, which deals with gravity. It marks a break, not only with the classical physics of Newton, with its absolute mechanical universe, but with the equally absolute classical geometry of Euclid. Einstein showed that Euclidean geometry only applied to "empty space," an ideally-conceived abstraction. In reality, space is not "empty." Space is inseparable from matter. Einstein maintained that space itself is conditioned by the presence of material bodies. In his general theory, this idea is conveyed by the seemingly paradoxical assertion that, near heavy bodies, "space is curved."

The real, i.e., material, universe is not at all like the world of Euclidean geometry, with the perfect circles, absolutely straight lines, and so on. The real world is full of irregularities. It is not straight, but precisely "warped." On the other hand, space is not something which exists separate and apart from matter. The curvature of space is just another way of expressing the curvature of matter which "fills" space. For example, it has been proved that light rays bend under the influence of the gravitational fields of bodies in space.

The general theory of relativity is essentially of a geometrical character, but this geometry is completely different to the classical Euclidean kind. In Euclidean geometry, for instance, parallel lines never meet or diverge, and the angles of a triangle always add up to 180°. Einstein's space-time (actually first developed by the Russian-German mathematician, Hermann Minkowski, one of Einstein's teachers, in 1907) represents a synthesis of three dimensional space (height, breadth and length) with time. This four-dimensional geometry deals with curved surfaces ("curved space-time"). Here the angles of a triangle may not add up to 180°, and parallel lines can cross or diverge.

In Euclidean geometry, as Engels points out, we meet a whole series of abstractions which do not at all correspond to the real world: a dimensionless point which becomes a straight line, which, in turn, becomes a perfectly flat surface, and so on and so forth. Among all these abstractions we have the emptiest abstraction of all, that of "empty space." Space, in spite of what Kant believed, cannot exist without something to fill it, and that something is precisely matter (and energy, which is the same thing). The geometry of space is determined by the matter which it contains. That is the real meaning of "curved space." It is merely a way of expressing the real properties of matter. The issue is only confused by inappropriate metaphors contained in popularizations of Einstein: "Think of space as a rubber sheet," or "Think of space as glass," and so on. In reality, the idea that must be kept in mind at all times is the indissoluble unity of time, space, matter and motion. The moment this unity is forgotten, we instantly slide into idealist mystification.

If we conceive space as a Thing-in-Itself, empty space, as in Euclid, clearly it cannot be curved. It is "nothing." However, as Hegel put it, there is nothing in the universe which does not contain both being and not-being. Space and matter are not two diametrically opposed, mutually exclusive phenomena. Space contains matter, and matter contains space. They are completely inseparable. The dialectical unity of matter and space is precisely what the universe is. In a most profound way, the general theory of relativity conveys this dialectical idea of the unity of space and matter. In the same way in mathematics zero itself is not "nothing," but expresses a real quantity, and plays a determining role.

Einstein presents gravitation as a property of space rather than a "force" acting upon bodies. According to this view space itself curves as a result of the presence of matter. This is a rather singular way of expressing the unity of space and matter, and one that is open to serious misinterpretations. Space itself, of course, cannot curve if it is understood as "empty space." The point is that it is impossible to conceive of space without matter. It is an inseparable unity. What we are considering is a definite relationship of space to matter. The Greek atomists long ago pointed out that atoms existed in the "void." The two things cannot exist without each other. Matter without space is the same as space without matter. A totally empty void is just nothing. But so is matter without any boundaries. Space and matter, then, are opposites which presuppose each other, define each other, limit each other, and cannot exist without each other.

The general theory served to explain at least one phenomenon which could not be explained by Newton's classical theory. As the planet Mercury approaches its closest point to the sun, its revolutions display a peculiar irregularity, which had been previously attributed to the perturbations caused by the gravity of other planets. However, even when these were taken into account, it did not explain the phenomenon. The deviation of Mercury's orbit around the sun ("perihelion") was very small, but enough to upset the astronomers' calculations. Einstein's general theory predicted that the perihelion of any revolving body should have a motion beyond that prescribed by Newton's law. This was shown to be correct for Mercury, and later also for Venus.

He also predicted that a gravitational field would bend light-rays. Thus, he claimed, a light ray passing close to the surface of the sun would be bent out of a straight line by 1.75 seconds of arc. In 1919 an astronomic observation of an eclipse of the sun showed this to be correct. Einstein's brilliant theory was demonstrated in practice. It was able to explain the apparent shift in the position of stars near the sun by the bending of their rays, and also the irregular motion of the planet Mercury, which could not be accounted for by Newton's theories.

Newton worked out the laws governing the movement of objects, according to which the strength of gravitational pull depends upon mass. He also maintained that any force exerted upon an object produces acceleration in inverse proportion to the mass of the object. Resistance to acceleration is called inertia. All masses are measured either through gravitational effects or inertial effects. Direct observation has shown that inertial mass and gravitational mass are, in fact, identical to within one part in one trillion. Einstein began his theory of general relativity by assuming that inertial mass and gravitational mass are exactly equal, because they are essentially the same thing.

The apparently motionless stars are moving at colossal speeds. Einstein's cosmic equations of 1917 implied that the universe itself was not fixed for all time, but could be expanding. The galaxies are moving away from us at speeds of about 700 miles a second. The stars and galaxies are constantly changing, coming into being and passing away. The whole universe is a vast arena where the drama of birth and death of stars and galaxies is played out across eternity. These are truly revolutionary events! Exploding galaxies, supernovas, catastrophic collisions between stars, black holes with a density billions of times greater than our sun greedily devouring entire clusters of stars. These things put in the shade the imaginings of the poets.

Relations Between Things

Many notions are purely relative in character. For example, if one is asked to say whether a road is on the right or left side of a house, it is impossible to answer. It depends on which direction one is moving relative to the house. On the other hand, it is possible to speak of the right bank of a river, because the current determines the direction of the river. Similarly, we can say that cars keep to the left (at least in Britain!) because the movement of a car singles out one of the two possible directions along the road. In all these examples, however, the notions "left" and "right" are shown to be relative, since they only acquire meaning after the direction by which they are defined as indicated.

In the same way, if we ask "Is it night or day?" the answer will depend on where we are. In London it is day, but in Australia it is night. Day and night are relative notions, determined by our position on the globe. An object will appear bigger or smaller depending upon its distance from a given point of observation. "Up" and "down" are also relative notions, which changed when it was discovered that the world is round, not flat. Even to this day, it is hard for "common sense" to accept that people in Australia can walk "upside down." Yet there is no contradiction if we understand that the notion of the vertical is not absolute but relative. For all practical purposes, we can take the earth's surface to be "flat" and therefore all verticals to be parallel, when dealing for instance, with two houses in one town. But when dealing with far larger distances, involving the whole earth's surface, we find that the attempt to make use of an absolute vertical leads to absurdities and contradictions.

By extension, the position of a planetary body is necessarily relative to the position of others. It is impossible to fix the position of an object without reference to other objects. The notion of "displacement" of a body in space means no more than that it changed its position relative to other bodies. A number of important laws of nature have a relativistic character, for example the principle of the relativity of motion and the law of inertia. The latter sates that an object on which no external force acts can only be not only in a state of rest but also in a state of uniform straight line motion. This fundamental law of physics was discovered by Galileo.

In practice, we know that objects upon which no external force is applied tend to come to rest, at least in everyday life. In the real world, the conditions for the law of inertia to apply, namely the total absence of external forces acting on the body, cannot exist. Forces such as friction act on the body to bring it to a halt. However, by constantly improving the condition of the experiment, it is possible to get closer and closer to the ideal conditions envisaged by the law of inertia, and thus show that it is valid even for the motions observed in everyday life. The relative (quantitative) aspect of time was perfectly expressed in Einstein's theories, which conveyed it far more profoundly than the classical theories of Newton.

Gravity is not a "force," but a relation between real objects. To a man falling off a high building, it seems that the ground is "rushing towards him." From the standpoint of relativity, that observation is not wrong. Only if we adopt the mechanistic and one-sided concept of "force" do we view this process as the earth's gravity pulling the man downwards, instead of seeing that it is precisely the interaction of two bodies upon each other. For "normal" conditions, Newton's theory of gravity agrees with Einstein's. But in extreme conditions, they completely disagree. In effect, Newton's theory is contradicted by the general theory of relativity in the same way as formal logic is contradicted by dialectics. And, to date, the evidence shows that both relativity and dialectics are correct.

As Hegel explained, every measurement is really the statement of a ratio. However, since every measurement is really a comparison, there must be one standard which cannot be compared with anything but itself. In general, we can only understand things by comparing them to other things. This expresses the dialectical concept of universal interconnections. To analyze things in their movement, development and relationships is precisely the essence of the dialectical method. It is the exact antithesis of the mechanical mode of thought (the "metaphysical" method in the sense of the word used by Marx and Engels) which views things as static and absolute. This was precisely the defect of the old classical Newtonian view of the universe, which, for all its achievements, never escaped from the one-sidedness which characterized the mechanistic world outlook.

The properties of a thing are not the result of relations to other things, but can only manifest themselves in relations to other things. Hegel refers to these relations in general as "reflex-categories." The concept of relativity is an important one, and was already fully developed by Hegel in the first volume of his masterpiece *The Science of Logic*.

We see this, for example, in social institutions such as kingship.

> Naïve minds [Trotsky observed], think that the office of kingship lodges in the king himself, in his ermine cloak and his crown, in his flesh and bones. As a matter of fact, the office of kingship is an interrelation between people. The king is king only because the interests and prejudices of millions of people are refracted through his person. When the flood of development sweeps away these interrelations, then the king appears to be only a washed-out man with a flabby lower lip. He who was once called Alfonso XIII could discourse upon this from fresh impressions.

> The leader by will of the people differs from the leader by will of God in that the former is compelled to clear the road for himself or, at any rate, to assist the conjuncture of events in discovering him. Nevertheless, the leader is always a relation between people, the individual supply to meet the collective demand. The controversy over Hitler's personality becomes the sharper the more the secret of his success is sought in himself. In the meantime, another political figure would be difficult to find that is in the same measure the focus of anonymous historic forces. Not every exasperated petty bourgeois could have become Hitler, but a particle of Hitler is lodged in every exasperated petty bourgeois.[37]

In *Capital*, Marx shown how concrete human labor becomes the medium for expressing abstract human labor. It is the form under which its opposite, abstract human labor, manifests itself. Value is not a material thing which can be derived from the physical properties of a commodity. In fact, it is an abstraction of the mind. But it is not on that account an arbitrary invention. In fact, it is an expression of an objective process, and is determined by the amount of socially necessary labor power expended in production. In the same way, time is an abstraction which, although it cannot be seen, heard or touched, and can only be expressed in relative terms as measurement, nevertheless denotes an objective physical process.

Space and time are abstractions which enable us to measure and understand the material world. All measurement is related to space and time. Gravity, chemical properties, sound, light, are all analyzed from these two points of view. Thus, the speed of light is 186,000 feet per second, while sound is determined by the number of vibrations per second. The sound of a stringed instrument, for instance, is determined by the time in which a certain number of vibrations occur and the spatial elements (length and thickness) of the vibrating body. That harmony which appeals to the aesthetic feelings of the mind is also another manifestation of ratio, measurement and therefore time.

Time cannot be expressed except in a relative way. In the same way, the magnitude value of a commodity can only be expressed relative to other commodities. Yet value is intrinsic to commodities, and time is an objective feature of matter in general. The idea that time itself is merely subjective, that is to say an illusion of the human mind, is reminiscent of the prejudice that money is merely a symbol, with no objective significance. The attempt to "demonetize" gold, which flowed from this false premise, led to inflation every time it was attempted. In the Roman Empire, the value of money was fixed by imperial decree, and it was forbidden to treat money as a commodity. The result was a continuous debasement of the currency. A similar phenomenon has taken place in modern capitalism, particularly since the Second World War. In economics, as in cosmology, the confusion of measurement with the nature of the thing itself leads to disaster in practice.

The Measurement of Time

While defining what time is presents a difficulty, measuring it does not. Scientists themselves do not explain what time is, but confine themselves to the measurement of time. From the mixing up of these two concepts endless confusion arises. Thus, Feynman:

> Maybe it is just as well if we face the fact that time is one of the things we cannot define (in the dictionary sense), and just say that it is what we already know it to be: it is how long we wait! What really matters anyway is not how we define time, but how we measure it. [38]

The measurement of time necessarily involves a frame of reference, and any phenomenon which entails change with time — e.g., the rotation of the earth or the swing of a pendulum. The earth's daily rotation on its axis provides a time scale. The decay of radioactive elements can be used for measuring long time intervals. The measurement of time involves a subjective element. The Egyptians divided day and night into twelfths. The Sumerians had a numerical system based on 60, and thus divided the hour into 60 minutes and the minute into 60 seconds. The meter was defined as one 10 millionth of the distance from the earth's pole to the equator (although this is not strictly accurate). The centimeter is 100th of a meter, and so on. At the beginning of this century, the investigation of the subatomic world led to the discovery of two natural units of measurement: the speed of light, c, and Planck's constant, h. These are not directly mass, length, or time, but the unity of all three.

There is an international agreement that the meter is defined as the distance between two scratches on a bar kept in a laboratory in France. More recently, it has been realized that this definition is neither as precise as would be useful, nor as permanent or universal as one would like. It is currently being considered that a new definition be adopted, an agreed-upon (arbitrary) number of wavelengths of a chosen spectral line. On the other hand, the measurement of time varies according to the scale and life-span of the objects under consideration.

It is clear that the concept of time will vary according to the frame of reference. A year on earth is not the same as a year on Jupiter. Nor is the idea of time and space the same for a human being as for a mosquito with a life-span of a few days, or a subatomic particle with a life span of a trillionth of a second (assuming, of course, that such entities could possess a concept of anything at all). What we are referring to here is the way time is perceived in different contexts. If we accept the given frame of references the way in which time would be seen would be different. Even in practice this can be seen, to some extent. For example, normal methods of measuring time cannot be applied to the measurement of the life-span of subatomic particles, and different standards must also be used for measuring "geological time."

From this point of view, time can be said to be relative. Measurement necessarily involves relationships. Human thought contains many concepts which are essentially relative, for example relative magnitudes, such as "big" and "small." A man is small compared to an elephant, but big in comparison to an ant. Smallness and bigness, in themselves, have no meaning. A millionth of a second, in ordinary terms, seems a very short length of time, yet at the subatomic level it is an extremely long time. At the other extreme, a million years is an extremely short time on the cosmological level.

All ideas of space, time and motion depend on our observations of the relations and changes in the material world. However, the measurement of time varies considerably when we consider different kinds of matter. The measurement of space and time is inevitably relative to some frame of reference — the earth, the sun, or any other static point — to which events of the universe can relate. Now it is clear that matter undergoes all kinds of different change: change of position, which, in turn, involves different velocities, change of state, involving different energy states, birth, decay and death, organization and disorganization, and many other transformations, all of which can be expressed and measured in terms of time.

In Einstein, time and space are not regarded as isolated phenomena, and indeed it is impossible to regard them as "things in themselves." Einstein advanced the view that time depends on the movement of a system and that the intervals of time change in such a way that the speed of light in the given system does not vary according to the movement. Spatial scales are also subject to change. The old classical Newtonian theories are still valid for everyday purposes, and even as a good approximation of the general workings of the universe. Newtonian mechanics still applies in a very wide branch of sciences, not only astronomy, but also practical sciences such as engineering. At low speeds, the effects of special relativity can be ignored. For example, the error involved in considering the behavior of a plane moving at 250 mile an hour would be about ten billionth of one percent. However, beyond certain limits it breaks down. At the kind of speeds that we find in particle acceleration, for example, it is necessary to take into account Einstein's prediction that mass is not constant, but increases with velocity.

From the point of view of our normal everyday notion of the measurement of time, the extremely short life-span of certain subatomic particles cannot be adequately expressed. A pi-meson, for instance, has a life-span of only about 10–16 of a second, before it disintegrates. Likewise, the period of a nuclear vibration, or the life-time of a strange resonance particle, is 10–24 second, approximately the time needed for light to cross the nucleus of a hydrogen atom. Another scale of measurement is necessary. Very short times, say 10–12 second, are measured by an electron beam oscilloscope. Even shorter times can be calibrated by means of laser techniques. At the other end of the scale, very long periods can be measured by a radioactive "clock."

In a sense, every atom in the universe is a clock, because it absorbs light (that is, electromagnetic rays) and emits it at precisely defined frequencies. Since 1967, the official internationally recognized standard of time is based on the atomic (caesium) clock. One second is defined as 9,192,631,770 vibrations of the microwave radiation from caesium-133 atoms during a specified atomic rearrangement. Even this highly accurate clock is not absolutely perfect. Different readings are taken from atomic clocks in about 80 different countries, and agreement is reached, "weighting" the time in favor of the steadiest clocks. By such means it is possible to arrive at accurate time-measurement to one millionth of a second per day, or even less.

For everyday purposes, "normal" time keeping, based on the rotation of the earth and the apparent movements of the sun and stars, is sufficient. But for a whole series of operations in the field of modern advanced technology, such as certain radio navigational aids in ships and airplanes, it becomes inadequate, leading to serious errors. It is at these kind of levels that the effects of relativity begin to make themselves felt. Experiments have shown that atomic clocks run slower at ground level than at high altitudes, where the gravitational effect is weaker. Atomic clocks, flown at an altitude of 30,000 feet, gained about three billionth of a second an hour. This conforms to Einstein's prediction to within one percent.

Problem Not Resolved

The special theory of relativity was one of the greatest achievements of science. It has revolutionized the way we look at the universe to such an extent that it has been compared with the discovery that the earth is round. Gigantic strides forward have been made possible by the fact that relativity established a far more accurate method of measurement than the old Newtonian laws it partially displaced. The philosophical question of time has, however, not been removed by Einstein's theory of relativity. If anything, it is more acute than ever. That there is something subjective and even arbitrary in the measurement of time is evident, as we have already commented. But this does not lead to the conclusion that time is purely a subjective thing. Einstein's entire life was spent in the pursuit of the objective laws of nature. The question is whether the laws of nature, including time, are the same for everyone, regardless of the place in which they are and the speed at which they are moving. On this question, Einstein vacillated. At times, he seemed to accept it, but elsewhere he rejected it.

The objective processes of nature are not determined by whether they are observed or not. They exist in and for themselves. The universe, and therefore time, existed before there were human beings to observe it, and will continue to exist long after there are no humans to concern themselves about it. The material universe is eternal, infinite, and constantly changing. However, in order that human minds may grasp the infinite universe, it is necessary to translate it into finite terms, to analyze and quantify it, so that it can become a reality for us. The way we observe the universe does not change it (unless it involves physical processes which interfere with what is being observed). But the way it appears to us can indeed change. From our standpoint, the earth appears to be at rest. But to an astronaut flying past our planet, it seems to be hurtling past him at a great speed. Einstein, who seems to have had a very dry sense of humor, apparently once asked an astonished ticket inspector: "What time does Oxford stop at this train?"

Einstein was determined to re-write the laws of physics in such a way that the predictions would always be correct, irrespective of the motions of different bodies, or the "points of view" which derive from them. From the standpoint of relativity, steady motion on a straight line is indistinguishable from being at rest. When two objects pass each other at a constant speed, it is equally possible to say that A is passing B, or that B is passing A. Thus, we arrive at the apparent contradiction that the earth is both at rest and moving at the same time. In the example of the astronaut, "it has to be simultaneously correct to say that the earth has great energy of motion and no energy and motion; the astronaut's point of view is just as valid as the view of learned men on earth." [39]

Although it seems straightforward, the measurement of time nevertheless presents a problem, because the rate of change of time must be compared to something else. If there is some absolute time, then this in turn must flow, and therefore must be measured against some other time, and so on ad infinitum. It is important to realize, however, that this problem presents itself only in relation to the measurement of time. The philosophical question of the nature of time itself does not enter into it. For the practical purposes of calculation and measurement, it is essential that a specific frame of reference by defined. We must know the position of the observer relative to the observed phenomena. Relativity theory shows that such statement as "at one and the same place" and "at one and the same time" are, in fact, meaningless.

The theory of relativity involves a contradiction. It implies that simultaneity is relative to a frame of axes. If one frame of axes is moving relative to another, then events that are simultaneous relative to the first are not simultaneous relative to the second, and vice versa. This fact, which flies in the face of common sense, has been experimentally demonstrated. Unfortunately, it can lend itself to an idealist interpretation of time, for instance, the assertion that there can be a variety of "presents." Moreover, the future can be portrayed as things and processes "that come into being" as four-dimensional solids that have as earliest temporal cross section or "time slice."

Unless this question is settled, all kinds of mistakes can be made: for example, the idea that the future already exists, and suddenly materializes in the "now," as a submerged rock suddenly appears when a wave breaks over it. In point of fact, both the past and the future are combined in the present. The future is being-in-potential. The past is what has already been. The "now" is the unity of both. It is actual being as opposed to potential being. Precisely for this reason, it is usual to feel regret for the past and fear for the future, not vice versa. The feeling of regret flows from the realization, corroborated by all human experience, that the past is lost forever, whereas the future is uncertain, consisting in a great number of potential states.

Benjamin Franklin once observed that there are only two things certain in this life - death and taxes, and the Germans have a proverb: "Man muss nur sterben" — "one only has to die," meaning that everything else is optional. Of course, this is not actually true. Many more things are inevitable than death, or even taxes. Out of an infinitely large number of potential states, in practice we know that only a certain number are really possible. Out of these, fewer still are probable at a given moment. And of the latter, in the end, only one will actually arise. The exact way in which this process unfolds is precisely the task of the different sciences to uncover. But this task will prove to be impossible if we do not accept that events and processes unfold in time, and that time is an objective phenomenon which expresses the most fundamental fact of all forms of matter and energy — change.

The material world is in a constant state of change, and therefore it "is and is not." This is the fundamental proposition of dialectics. Philosophers like the Anglo-American Alfred North Whitehead and the French intuitionist Henry Begson believed that the flow of time was a metaphysical fact which could only be grasped by non-scientific intuition. "Process philosophers" like these, despite their mystical overtones, at least are correct in saying that the future is open or indeterminate whereas the past is unchangeable, fixed and determinate. It is "congealed time." On the other hand we have the "philosophers of the manifold" who maintain that future events may exist but not be connected in a sufficiently law-like way with past events. Pursuing a philosophically incorrect view of time, we end up with sheer mysticism, as in the notion of the "multiverse" — an infinite number of "parallel" universes (if that is the right word, since they do not exist in space "as we know it") existing simultaneously (if that is the right word, since they do not exist in time "as we know it"). Such is the confusion that arises from the idealist interpretation of relativity.

Idealist Interpretations

> *There was a young lady named Bright*
> *Whose speed was faster than light;*
> *She set out one day*
> *In a relative way*
> *And returned home the previous night.*
> (A. Buller, *Punch*, 19th December 1923)

As with quantum mechanics, relativity has been seized upon by those who wish to introduce mysticism into science. "Relativity" is taken to mean that we cannot really know the world. As J. D. Bernal explains:

> It is, however, equally true that the effect of Einstein's work, outside the narrow specialist fields where it can be applied, was one of general mystification. It was eagerly seized on by the disillusioned intellectuals after the First World War to help them in refusing to face realities. They only needed to use the word "relativity" and say "Everything is relative," or "It depends on what you mean." [40]

This is a complete misinterpretation of Einstein's ideas. In point of fact, the very word "relativity" is a misnomer. Einstein himself preferred the name invariance theory which gives a far better idea of what he intended — the exact opposite of the vulgar idea of relativity theory. It is quite untrue that for Einstein "everything is relative." To begin with, rest energy (that is, the unity of matter and energy) is one of the absolutes of the theory of relativity. The limiting speed of light is another. Far from an arbitrary, subjective interpretation of reality, in which one opinion is as good as another, and "it all depends how you look at it," Einstein "discovered what was 'absolute' and reliable despite the apparent confusions, illusions and contradictions produced by relative motions or the action of gravity." [41]

The universe exists in a constant state of change. In that sense, nothing is "absolute" or eternal. The only absolute is motion and change, the basic mode of existence of matter — something which Einstein demonstrated conclusively in 1905. Time and space, as the mode of existence of matter are objective phenomena. They are not merely abstractions or arbitrary notions invented by humans (or gods) for their own convenience, but fundamental properties of matter, expressing the universality of matter.

Space is three dimensional, but time has only one dimension. With apologies to the makers of films in which it is possible to "go back to the future," it is only possible to travel in one direction in time, from the past to the future. There is no more danger of a spaceman returning to earth before he was born, or of a man marrying his great grandmother, than there is of any of the other amusing but idiotic fantasies of Hollywood. Time is irreversible, which is to say, every material process develops in only one direction — from the past to the future. Time is merely a way of expressing the real movement and changing state of matter. Matter, motion, time and space are inseparable.

The shortcoming of Newton's theory was to regard space and time as separate entities, one alongside the other, independent of matter and motion. Up till the 20th century, scientists identified space with a vacuum (a "nothing"), which was seen as something absolute, that is, always and everywhere the same, changeless "thing." These empty abstractions have been discredited by modern physics, which has demonstrated the profound relation between time, space, matter and motion. Einstein's relativity theory firmly establishes that time and space do not exist in and of themselves, in isolation from matter, but are part of a universal interrelation of phenomena. This is conveyed by the concept of the integral and indivisible space-time, of which time and space are seen as relative aspects. A controversial idea here is the prediction that a clock in motion will keep time more slowly than one that is stationary. However, it is important to understand that this effect only becomes noticeable at extraordinarily high speeds, approaching the speed of light.

If Einstein's general theory of relativity is correct, then the theoretical possibility would exist in the future of traveling unimaginable distances through space. Theoretically, it would be possible for a human being to survive thousands of years into the future. The whole question hinges upon whether the changes observed in rates of atomic clocks also apply to the rate of life itself. Under the effect of strong gravity, atomic clocks run slower than in empty space. The question is whether the complex interrelations of molecules which constitute life can behave in the same way. Isaac Asimov, who knew a thing or two about science fiction, wrote: "If time really slows down in motion, one might journey even to a distant star in one's own lifetime. But of course one would have to say good-bye to one's own generation and return to the world of the future." 42

The argument for this is that the rates of living processes are determined by the rates of atomic action. Thus, under strong gravity, the heart will beat more slowly, and the brain impulses will also slow down. In fact, all energy diminishes in the presence of gravity. If processes slow down, they also take longer in time. If a space-ship were able to travel close to the speed of light, the universe would be seen flashing past it, while for those inside, time would continue as "normal," i.e., at a much slower rate. The impression would be that time outside would be speeded up. Is that correct? Would he in fact be living in the future, relative to people on earth, or not? Einstein seems to answer in the affirmative.

All kinds of mystical notions arise from such speculation — for example about hopping into a black hole and entering another universe. If a black hole exists, and that is still not definitely proven, all that would be at the center would be the collapsed remains of a gigantic star, not another universe. Any real person who entered it would be instantly torn apart and converted into pure energy. If that is what is considered as passing into another universe, then those who advocate such ideas are most welcome to make the first excursion! In reality, this is pure speculation, however entertaining it may be. The whole idea of "time-travel" inevitably lands one in a mass of contradictions, not of dialectical but of the absurd variety. Einstein would have been shocked at the mystical interpretation of his theories which involve notions such as shuttling back and forth in time, altering the future, and nonsense of that sort. But he himself must bear some responsibility for this situation because of the idealist element in his outlook, particularly on the question of time.

Let us grant that an atomic clock at a high altitude runs faster at high altitudes than on the ground, because of the effect of gravitation. Let us also grant that, when this clock returns to earth, it is found to be, say, 50 billionth of a second older than equivalent clocks which had never left the ground. Does that mean that a man traveling in the same flight has equally aged? The process of ageing is dependent upon the rate of metabolism. This is partly influenced by gravitation, but also by many other factors. It is a complex biological process, and it is not easy to see how it could be fundamentally affected either by velocity or gravitation, except that extremes of either can cause material damage to living organisms.

If it were possible to slow down the rate of metabolism in the way predicted, so that, for example, the heart-beat would slow to one every twenty minutes, the process of ageing would presumably be correspondingly slower. It is, in fact, possible to slow down metabolism, for example, by freezing. Whether this would be the effect of traveling at very high speeds, without killing the organism is open to doubt. According to the well-known theory, such a relativistic space-man, if he succeeded on returning to earth, would come back after, say 10,000 years, and to pursue the usual analogy, would presumably be in a position to marry his own remote descendants. But he would never be able to return to his "own" time.

Experiments conducted with subatomic particles (muons) indicate that particles traveling at 99.94 per cent of the speed of light extended their life by nearly thirty times, precisely as predicted by Einstein. However, whether these conclusions can be applied to matter on a larger scale, and living matter in particular, is an issue which remains to be seen. Many serious mistakes have been made by attempting to apply the results derived from one sphere to another, entirely different, area. In the future, space-travel at very high speeds — maybe one-tenth of the speed of light — may become possible. At such speed, a journey of five light-years would take fifty years (though according to Einstein, it would take three months less for those traveling). Will it ever be possible to travel at the speed of light, thus enabling human beings to reach the stars? At this moment in time, such a prospect seems remote. But then, a hundred years ago — a mere blink in history — the idea of traveling to the moon was still confined to the novels of Jules Verne.

Mach and Positivism

> *The object, however, is the real truth, is the essential reality; it is, quite indifferent to whether it is known or not; it remains and stands even though it is not known, while the knowledge does not exist if the object is not there.* (Hegel) [43]

The existence of past, present and future is deeply engraved on the human consciousness. We live now, but we can remember past events, and, to some extent, foresee future ones. There is a "before" and an "after." Yet some philosophers and scientists dispute this. They regard time as a product of the mind, an illusion. In their view, in the absence of human observers, there is no time, no past, present or future. This is the standpoint of subjective idealism, an entirely irrational and anti-scientific outlook which nevertheless has attempted for the last hundred years to base itself in the discoveries of physics to lend respectability to what is essentially a mystical view of the world. It seems ironical that the school of philosophy which has had the biggest impact upon science in the 20th century, logical positivism, is precisely a branch of subjective idealism.

Positivism is a narrow view which holds that science should confine itself to the "observed facts." The founders of this school were reluctant to refer to theories as true or false, but preferred to describe them as more or less "useful." It is interesting to note that Ernst Mach, the real spiritual father of neo-positivism, opposed the atomist theory of physics and chemistry. This was the natural consequence of the narrow empiricism of the positivist outlook. Since the atom could not be seen, how could it exist? It was regarded by them at best as a convenient fiction, and at worst as an unacceptable ad hoc hypothesis. One of Mach's co-thinkers, Wilhelm Ostwald actually attempted to derive the basic laws of chemistry without the help of the atomic hypothesis!

Boltzmann sharply criticized Mach and the Positivists, as did Max Planck, the father of quantum physics. Lenin subjected the views of Mach and Richard Avenarius, the founder of the school of Empirio-criticism, to a devastating criticism in his book *Materialism and Empirio-criticism*, (1908). Nevertheless, the views of Mach had a big impact and, among others, impressed the young Albert Einstein. Setting out from the view of that all ideas must be derived from "the given," that is, from the information provided immediately by our senses, they went on to deny the existence of the natural world, independent of human sense-perception. Mach and Avenarius referred to physical objects as "complexes of sensation." Thus, for example, this table is no more than a collection of sense-impressions such as hardness, color, mass and so on. Without these, they maintained, nothing would be left. Therefore, the idea of matter (in the philosophical sense, that is, the objective world given to us in sense-perception) was declared to be meaningless.

As we have already pointed out, these ideas lead directly to solipsism — the idea that only "I" exist. If I close my eyes, the world ceases to exist. Mach attacked Newton's idea that space and time are absolute and real entities, but he did so from the standpoint of subjective idealism. Incredibly, the most influential school of modern philosophy (and the one that had the biggest influence on scientists) was derived from the subjective idealism of Mach and Avenarius.

The obsession with "the observer" which is a thread running through the whole of 20th century theoretical physics is derived from the subjective idealist philosophy of Ernst Mach. Taking his starting-point from the empiricist argument that "all our knowledge is derived from immediate sense-perception," Mach argued that objects cannot exist independently of our consciousness. Carried to its logical conclusion, this would mean that, for example, the world could not have existed before there were people present to observe it. As a matter of fact, it could not have existed before I was present, since I can only know my own sensations, and cannot therefore be sure that any other consciousness exists.

The important thing is that Einstein himself was initially impressed by these arguments, which left their mark on his early writings on relativity. This has, beyond doubt, exercised the most harmful influence upon modern science. Whereas Einstein was capable of realizing his mistake, and attempted to correct it, those who have slavishly followed the master, have been incapable of sorting out the chaff from the grain. As often happens, over-eager disciples become dogmatic. They are more Papist than the Pope! In his autobiography, Karl Popper shows clearly that in his later years Einstein regretted his earlier subjective idealism, or "operationalism," which demanded the presence of an observer to determine natural processes:

> It is an interesting fact that Einstein himself was for years a dogmatic positivist and operationalist. He later rejected this interpretation: he told me in 1950 that he regretted no mistake he ever made as much as this mistake. The mistake assumed a really serious form in his popular book, *Relativity: The Special and the General Theory*. There he says "I would ask the reader not to proceed farther until he is fully convinced on this point." The point is, briefly, that "simultaneity" must be defined — and defined in an operational way — since otherwise "I allow myself to be deceived...when I imagine that I am able to attach a meaning to the statement of simultaneity." Or in other words, a term has to be operationally defined or else it is meaningless. (Here in a nutshell is the positivism later developed by the Vienna Circle under the influence of Wittgenstein's *Tractatus*, and in a very dogmatic form).

This is important, because it shows that Einstein in the end rejected the subjectivist interpretation of relativity theory. All the nonsense about "the observer" as a determining factor was not an essential part of the theory, but merely the reflection of a philosophical mistake, as Einstein frankly confirmed. That, unfortunately, did not prevent the followers of Einstein from taking over the mistake, and blowing it up to the point where it appeared to be a fundamental cornerstone of relativity. It is here that we find the real origin of Heisenberg's subjective idealism:

> But many excellent physicists," Popper continues, "were greatly impressed by Einstein's operationalism, which they regarded (as did Einstein himself for a long time) as an integral part of relativity. And so it happened that operationalism became the inspiration of Heisenberg's paper of 1925, and of his widely accepted suggestion that the concept of the track of an electron, or of its classical position-cum-momentum, was meaningless. [44]

The fact that time is an objective phenomenon, reflecting real processes in nature was first demonstrated by the laws of thermodynamics, which were worked out in the 19th century and which still play a central role in modern physics. These laws, particularly as developed by Boltzmann, firmly establish the idea not only that time exists objectively, but that it flows in only one direction, from past to future. Time cannot be reversed, nor is it dependent upon any "observer."

Boltzmann and Time

The fundamental question which has to be addressed is: Is time an objective feature of the physical universe? Or is it something purely subjective, an illusion of the mind, or merely a convenient way of describing things to which it has no real relationship? The latter position has been held, in one or other degree, by a number of different schools of thought, all of them closely related to the philosophy of subjective idealism. Mach, as we have seen, introduced this subjectivism into science. It was decisively answered towards the end of the 19th century by the pioneer of the science of thermodynamics, Ludwig Boltzmann.

Einstein, under the influence of Ernst Mach, treated time as something subjective, which depended on the observer, at least in the beginning before he realized the harmful consequences of this approach. In 1905, his paper on the special theory of relativity introduced the notion of a "local time" associated with each separate observer. The concept of time here contains an idea carried over from classical physics, namely that time is reversible. This is really quite an extraordinary notion, and one which flies in the face of all experience. Film directors often resort to a trick photography, in which the camera is put into reverse, whereupon the most peculiar events occur: milk flows from the glass back into the bottle, buses and cars run backwards, eggs return to their shells, and so on. Our reaction to all this is to laugh, which is what is intended. We laugh because we know that what we are seeing is not just impossible, but absurdly so. We know that the processes which we are seeing cannot be reversed.

Boltzmann understood this, and the concept of irreversible time lies at the heart of his famous theory of the arrow of time. The laws of thermodynamics represented a major breakthrough in science, but were controversial. These laws could not be reconciled with the existing laws of physics at the end of the 19th century. The second law cannot be derived from the laws of mechanics or quantum mechanics, and, in effect, marks a sharp break with the theories of previous physical science. It says that entropy increases in the direction of the future, not the past. It denotes a change in state over time, which is irreversible. The notion of a tendency towards dissipation clashed with the accepted idea that the essential task of physics was to reduce the complexity of nature to simple laws of motion.

The idea of entropy, which is usually understood as a the tendency of things towards greater disorganization and decay with the passing of time, entirely bears out what people have always believed: that time exists objectively and that it is a one-way process. The two laws of thermodynamics imply the existence of the phenomenon known as entropy that is conserved in all irreversible processes. Its definition is based on another property known as available energy. The entropy of an isolated system may remain constant or increase, but it cannot decrease. One of the results of this is the impossibility of a "perpetual motion machine."

Einstein considered the idea of irreversible time to be an illusion that had no place in physics. In Max Planck's words, the second law of thermodynamics expresses the idea that there exists in nature a quantity which changes always in the same sense in all natural processes. This does not depend on the observer, but is an objective process. But Planck's view was in a small minority. The great majority of scientists, like Einstein, attributed it to subjective factors. Einstein's position on this question shows up the central weakness of his standpoint in making objective processes depend upon a non-existent "observer." This was undoubtedly the weakest element in his entire outlook, and, for that very reason, is the part which has proved most popular with his successors, who do not seem aware of the fact that Einstein himself changed his mind on this towards the end of his life.

In physics and mathematics the expression of time is reversible. A "time-reversal invariant" implies that the same laws of physics apply equally well in both situations. The second event is indistinguishable from the first and the flow of time does not have any preferred direction in the case of fundamental interactions. For example, a film of two billiard balls colliding can be run forward or backward, without giving any idea of the true time sequence of the event. The same was assumed to be true of interactions at the sub atomic level, but evidence to the contrary was found in 1964 in weak nuclear interactions. For a long time it was believed that the fundamental laws of nature were "charge symmetrical." For example, an antiproton and a positron behave like a proton and an electron. Experiments have now shown that the laws of nature are symmetrical if three basic things are combined — time, charge and parity. This is known as a "CPT mirror."

In dynamics, the direction of a given trajectory was irrelevant. For example, a ball bouncing on the ground would return to its initial position. Any system can thus "go backwards in time," if all the points involved in it are reversed. All the states it previously went through would simply be retraced. In classical dynamics, changes such as time reversal ($t \to -t$) and velocity reversal ($v \to -v$) are treated as mathematically equivalent. This kind of calculation works well for simple closed systems, where there are no interactions. In reality, however, every system is subject to many interactions. One of the most important problems in physics is the "three-body" problem, for example, the moon's motion is influenced by the sun and the earth. In classical dynamics, a system changes according to a trajectory that is given once and for all, the starting point of which is never forgotten. Initial conditions determine the trajectory for all time. The trajectories of classical physics were simple and deterministic. But there are other trajectories that are not so easy to pin down, for example, a rigid pendulum, where an infinitesimal disturbance would be enough to set it rotating or oscillating.

The importance of Boltzmann's work was that he dealt with the physics of processes rather than the physics of things. His greatest achievement was to show how the properties of atoms (mass, charge, structure) determine the visible properties of matter (viscosity, thermal conductivity, diffusion, etc.). His ideas were viciously attacked during his lifetime, but vindicated by the discoveries of atomic physics shortly before 1900, and the realization that the random movements of microscopic particles suspended in a fluid ("Brownian motion") could only be explained in terms of the statistical mechanics invented by Boltzmann.

The bell-shaped Gauss curve describes the random motion of molecules in a gas. An increased temperature leads to an increase in the average velocity of the molecules and the energy associated with their motion. Whereas Clausius and Maxwell approached this question from the standpoint of the trajectories of individual molecules, Boltzmann considered the population of molecules. His kinetic equations play an important role in the physics of gases. It was a major advance in the physics of processes. Boltzmann was a great pioneer, who was treated as a madman by the scientific establishment. He was finally driven to suicide in 1906, having previously been compelled to retreat from his attempt to establish the irreversible nature of time as an objective feature of nature.

Whereas in the theory of classical mechanics, the events in the film earlier described are perfectly possible, in practice, they are not. In the theory of dynamics, for example, we have an ideal world in which such things as friction and collision do not exist. In this ideal world, all the invariants involved in a given motion are fixed at the start. Nothing could happen to alter its course. By these means, we arrive at a completely static view of the universe, where everything is reduced to smooth, linear equations. Despite the revolutionary advances made possible by relativity theory, Einstein, at heart, remained wedded to the idea of a static, harmonious universe — just like Newton.

The equations of motion of Newtonian or for that matter quantum mechanics have no built-in irreversibility. It is possible to run a movie film forward or backwards. But this is not true of nature in general. The second law of thermodynamics predicts an irreversible tendency towards disorder. It states that randomness always increases in time. Until recently, it was thought that the fundamental laws of nature are symmetrical in time. Time is asymmetrical and moves only in one direction, from past to future. We see fossils, footprints, photographs and hear recordings of things from the past, but never from the future. It is easy to mix eggs to make an omelet or put milk and sugar into a cup of coffee, but not to reverse these processes. The water in the bath transfers its heat to the surrounding air, but not vice versa.

The second law of thermodynamics is the "arrow of time." The subjectivists objected that irreversible processes like chemical affinity, heat conduction, viscosity, etc., would depend on the "observer." In reality, they are objective processes that take place in nature, and this is clear to everyone in relation to life and death. A pendulum (at least in an ideal state) can swing back to its initial position. But everyone knows that the life of an individual moves in only one direction, from the cradle to the grave. It is an irreversible process. Ilya Prigogine, one of the leading theorists of chaos theory, has devoted a lot of attention to the question of time. When he first began to study physics as a student in Brussels, Prigogine recalls that he was "astonished by the fact that science had so little to say about time, especially since [his] earlier education had centered mainly around history and archaeology." In relation to the conflict between classical mechanics (dynamics) and thermodynamics, Prigogine and Stengers write:

> To a certain extent, there is an analogy between this conflict and the one that gave rise to dialectical materialism. We have described...a nature that might be called "historical" — that is, capable of development and innovation. The idea of a history of nature as an integral part of materialism was asserted by Marx and, in greater detail, by Engels. Contemporary developments in physics, the discovery of the constructive role played by irreversibility, have thus raised within the natural sciences a question that has long been asked by materialists. For them, understanding nature meant understanding it as being capable of producing man and his societies.
>
> Moreover, at the time Engels wrote his *Dialectics of Nature*, the physical sciences seemed to have rejected the mechanistic world view and drawn close to the idea of an historical development of nature. Engels mentions three fundamental discoveries: energy and the laws governing its qualitative transformations, the cell as the basic constituent of life, and Darwin's discovery of the evolution of species. In view of these great discoveries, Engels came to the conclusion that the mechanistic world view was dead."

Against the subjective interpretation of time, the authors conclude: "Time flows in a single direction, from past to future. We cannot manipulate time, we cannot travel back to the past." [45]

Relativity and Black Holes

In Einstein's view, unlike that of Newton, gravity affects time because it affects light. If one could imagine a particle of light poised on the edge of a black hole, it would be suspended indefinitely, neither advancing nor retreating, neither losing energy, nor gaining it. In such a state, it is possible to argue that "time stands still." This is the argument of the relativist proponents of the black hole and its properties. What this boils down to is that if all motion were to cease, then there would be no change either of state or position, and therefore no time in any meaningful sense of the word. Such a situation is alleged to exist at the edge of a black hole. This, however, seems a highly speculative and mystical interpretation of a phenomenon, the existence of which is, in itself, unproved.

All matter exists in a constant state of change and motion, and therefore, all that is being said here is that if matter and motion are eliminated, there is no time either, which is a complete tautology. It is like saying — if there is no matter, there is no matter, or if there is no time, there is no time. Because both statements mean just the same thing. Strangely enough, one would seek in vain in the theory of relativity for a definition of what time and space are. Einstein certainly found it difficult to explain. However, he came close to it when he explained the difference between his geometry and the classical geometry of Euclid. He said that one could imagine a universe in which space was not warped, but it would be completely devoid of matter. This points clearly in the right direction. After all the fuss about black holes, you may also be surprised to discover that this subject was not even mentioned by Einstein. He relied upon a rigorous approach, mainly based on very complicated mathematics, and made predictions which could be verified by observation and experiment. The physics of black holes, in the absence of clearly established empirical data, has an extremely speculative character.

Despite its successes, it is still possible that the general theory of relativity may be wrong. Unlike special relativity, the experimental tests which have been carried out on it are not very many. There is no conclusive proof, although up to the present time no conflict has been found between the theory and the observed facts. It is not even ruled out that the assertion of special relativity, that nothing can move faster than the speed of light, may be shown to be incorrect in the future.*

Alternative theories of relativity have been put forward, for example, by Robert Dicke. Dicke's theory predicted a deflection of the moon's orbit of several feet towards the sun. Using advanced laser technology, the McDonald observatory in Texas found no trace of this displacement. However, there is no reason to suppose that the last word has been spoken. So far, Einstein's theories have been borne out by repeated experiment. But the constant probing of extreme conditions must sooner or later reveal a set of circumstances that are not covered by the equations, preparing the way for new epoch-making discoveries. The theory of relativity cannot be the end of the line, any more than Newtonian mechanics, Maxwell's theory of electromagnetism, or any previous theory.

For two hundred years, the theories of Newton were held to be absolutely valid. His authority could not be challenged. After his death, Laplace and others carried his theories to an extreme where they became absurd. The radical break with the old mechanistic Absolutes was a necessary condition for the further advance of physics in the 20th century. It was the proud boast of the new physics that they had forever killed off the ogre of the Absolute. Suddenly thought was free to move into hitherto unheard of realms. These were heady times! Unfortunately, such happiness cannot last forever. In the words of Robert Burns:

> But pleasures are like poppies spread:
> You seize the flow'r, its bloom is shed.

The new physics solved many problems, but only at the cost of creating new contradictions, which remain unresolved even at the present time. For most of the present century, physics has been dominated by two imposing theories: quantum mechanics and relativity. What is not generally realized is that the two theories are at variance. In fact, they are incompatible. The general theory of relativity takes no account whatever of the uncertainty principle. Einstein spent the last years of his life attempting to resolve this contradiction, but failed to do so.

Relativity theory was a great and revolutionary theory. So was Newtonian mechanics in its day. Yet it is the fate of all such theories to become transformed into orthodoxies, to suffer a kind of hardening of the arteries, until they are no longer able to answer the questions posed by the march of science. For a long time, theoretical physicists have been content to rest on the discoveries of Einstein, in the same way that an earlier generation were content to swear by Newton. And in just the same way, they are guilty of bringing general relativity into disrepute by reading into it the most absurd and fantastic notions, which its author never even dreamed of.

Singularities, black holes where time stands still, multiverses, a time before time began, about which no questions must be asked — one can imagine Einstein clutching his head! All this is supposed to flow inevitably from general relativity, and anyone who raised the slightest doubt about it is immediately confronted with the authority of the great Einstein. This is not one whit better than the situation before relativity, when the authority of Newton was similarly wielded in defense of the existing orthodoxy. The only difference is that the fantastic notions of Laplace seem extremely sensible alongside the mystical gobbledygook written by some physicists today. And even less than Newton can Einstein be made responsible for the outlandish flights of fancy of his successors, which represent the reductio ad absurdum of the original theory.

These senseless and arbitrary speculations are the best proof that the theoretical framework of modern physics is in need of a complete overhaul. For the problem here is one of method. It is not just that they provide no answers. The problem is that they do not even know how to ask the right questions. This is not so much a scientific as a philosophical question. If everything is possible, then one arbitrary theory (more correctly, guess) is as good as the next. The whole system has been pushed near to breaking-point. And to cover up the fact, they resort to a mystical kind of language, in which the obscurity of expression does not disguise the complete lack of any real content.

This state of affairs is clearly intolerable, and has led a section of scientists to begin to question the basic assumptions on which science has been operating. David Bohm's investigations into the theory of quantum mechanics, Ilya Prigogine's new interpretation of the Second Law of Thermodynamics, Hannes Alfvén's attempt to work out an alternative to the orthodox cosmology of the big bang, above all, the spectacular rise of chaos and complexity theory — all this indicates the existence of a ferment in science. While it is too early to predict the exact outcome of this, it seems likely that we are entering into one of those exciting periods in the history of science, when an entirely new approach will emerge.

There is every reason to suppose that eventually the theories of Einstein will be surpassed by a new and broader-based theory, which, while preserving all that is viable in relativity, will correct and amplify it. In the process, we shall certainly arrive at a truer and more balanced understanding of the questions relating to the nature of time, space and causality. This does not signify a return to the old mechanical physics, any more than the fact that we can now achieve the transformation of the elements means a return to the ideas of the alchemists. As we have seen, the history of science frequently involves an apparent return to earlier positions, but on a qualitatively higher level.

One thing we can predict with absolute confidence: when the new physics finally emerges from the present chaos there will be no place in it for time-travel, multiverses, or singularities which compress the whole of the universe into a single point, about which no questions are allowed to be asked. This will sadly make it much more difficult to win big cash prizes for providing the Almighty with scientific credentials, a fact which some may regret, but which, in the long term, may not be a bad thing for the progress of science!

* This prediction appears to have been confirmed far sooner than we expected. Before the book was sent to the printers, reports appeared in the press of an experiment conducted by American scientists which appear to indicate that photons can travel faster than the speed of light. The experiment is a complicated one, based on a peculiar phenomenon known as "quantum tunneling." If it is shown to be correct, this will demand a fundamental rethinking of the whole concept of relativity.

8.   THE ARROW OF TIME

The Second Law of Thermodynamics

> *This is the way the world ends*
> *Not with a bang but a whimper.* (T. S. Eliot)

Thermodynamics is the branch of theoretical physics which deals with the laws of heat motion, and the conversion of heat into other types of energy. The word is derived from the Greek words *therme* ("heat") and *dynamis* ("force"). It is based upon two fundamental principles originally derived from experiments, but which are now regarded as axioms. The first principle is the law of the conservation of energy, which assumes the form of the law of the equivalence of heat and work. The second principle states that heat cannot of itself pass from a cooler body to a hotter body without changes in any other bodies.

The science of thermodynamics was a product of the industrial revolution. At the beginning of the 19th century, it was discovered that energy can be transformed in different ways, but can never be created or destroyed. This is the first law of thermodynamics — one of the fundamental laws of physics. Then, in 1850, Robert Clausius discovered the second law of thermodynamics. This states that "entropy" (i.e., the ratio of a body's energy to its temperature) always increases in any transformation of energy, for example, in a steam engine.

Entropy is generally understood to signify an inherent tendency towards disorganization. Every family is well aware that a house, without some conscious intervention, tends to pass from a state of order to disorder, especially when young children are around. Iron rusts, wood rots, dead flesh decays, the water in the bath gets cold. In other words, there appears to be a general tendency towards decay. According to the second law, atoms, when left to themselves, will mix and randomize themselves as much as possible. Rust occurs because the iron atoms tend to mingle with oxygen in the surrounding air to form iron oxide. The fast moving molecules on the surface of the bath water collide with the slower moving molecules in the cold air and transfer their energy to them.

This is a limited law, which has no bearing on systems consisting of a small number of particles (microsystems) or to systems with an infinitely large number of particles (the universe). However, there have been repeated attempts to extend its application well beyond the proper sphere, leading to all kinds of false philosophical conclusions. In the middle of the last century, R. Clausius and W. Thomson, the authors of the second principle of thermodynamics, attempted to apply the second law to the universe as a whole, and arrived at a completely false theory, known as the "thermal death" theory of the end of the universe.

This law was redefined in 1877 by Ludwig Boltzmann, who attempted to derive the second law of thermodynamics from the atomic theory of matter, which was then gaining ground. In Boltzmann's version, entropy appears as a function of the probability of a given state of matter: the more probable the state, the higher its entropy. In this version, all systems tend towards a state of equilibrium (a state in which there is no net flow of energy). Thus, if a hot object is placed next to a cold one, energy (heat) will flow from the hot to the cold, until they reach equilibrium, i.e., they both have the same temperature.

Boltzmann was the first one to deal with the problems of the transition from the microscopic (small-scale) to the macroscopic (large-scale) level in physics. He attempted to reconcile the new theories of thermodynamics with the classical physics of trajectories. Following Maxwell's example, he tried to resolve the problems through the theory of probability. This represented a radical break with the old Newtonian methods of mechanistic determinism. Boltzmann realized that the irreversible increase in entropy could be seen as the expression of a growing molecular disorder. His principle of order implies that the more probable state available to a system is one in which a multiplicity of events taking place simultaneously within the system cancel each other out statistically. While molecules can move randomly, on average, at any given moment, the same number will be moving in one direction as in another.

There is a contradiction between energy and entropy. The unstable equilibrium between the two is determined by temperature. At low temperatures, energy dominates and we see the emergence of ordered (weak-entropy) and low energy states, as in crystals, where molecules are locked in a certain position relative to other molecules. However, at high temperature, entropy prevails, and is expressed in molecular disorder. The structure of the crystal is disrupted, and we get the transition, first to a liquid, then to a gaseous state.

The second law states that the entropy of an isolated system always increases, and that when two systems are joined together, the entropy of the combined system is greater than the sum of the entropies of the individual systems. However, the second law of thermodynamics is not like other laws of physics, such as Newton's law of gravity, precisely because it is not always applicable. Originally derived from a particular sphere of classical mechanics, the second law is limited by the fact that Boltzmann took no account of such forces as electromagnetism or even gravity, allowing only for atomic collisions. This gives such a restricted picture of physical processes, that it cannot be taken as generally applicable, although it does apply to limited systems, like boilers. The Second Law is not true of all circumstances. Brownian motion contradicts it, for example. As a general law of the universe in its classical form, it is simply not true.

It has been claimed that the second law means that the universe as a whole must tend inexorably towards a state of entropy. By an analogy with a closed system, the entire universe must eventually end up in a state of equilibrium, with the same temperature everywhere. The stars will run out of fuel. All life will cease. The universe will slowly peter out in a featureless expanse of nothingness. It will suffer a "heat-death." This bleak view of the universe is in direct contradiction to everything we know about its past evolution, or see at present. The very notion that matter tends to some absolute state of equilibrium runs counter to nature itself. It is a lifeless, abstract view of the universe. At present, the universe is very far from being in any sort of equilibrium, and there is not the slightest indication either that such a state ever existed in the past, or will do so in the future. Moreover, if the tendency towards increasing entropy is permanent and linear, it is not clear why the universe has not long ago ended up in a tepid soup of undifferentiated particles.

This is yet another example of what happens when attempts are made to extend scientific theories beyond the limits where they have a clearly proven application. The limitations of the principles of thermodynamics were already shown in the last century in a polemic between Lord Kelvin, the celebrated English physicist, and geologists, concerning the age of the earth. The predictions made by Lord Kelvin on the basis of thermodynamics ran counter to all that was known by geological and biological evolution. The theory postulated that the earth must have been molten just 20 million years ago. A vast accumulation of evidence proved the geologists right, and Lord Kelvin wrong.

In 1928, Sir James Jean, the English scientist and idealist, revived the old arguments about the "heat death" of the universe, adding in elements taken from Einstein's relativity theory. Since matter and energy are equivalents, he claimed, the universe must finally end up in the complete conversion of matter into energy: "The second law of thermodynamics," he prophesied darkly, "compels materials in the universe [sic!] to move ever in the same direction along the same road which ends only in death and annihilation." [46]

Similar pessimistic scenarios have been put forward more recently. In the words of a book, published recently:

> The universe of the very far future would thus be an inconceivably dilute soup of photons, neutrinos, and a dwindling number of electrons and positrons, all slowly moving farther and farther apart. As far as we know, no further basic physical processes would ever happen. No significant event would occur to interrupt the bleak sterility of a universe that has run its course yet still faces eternal life — perhaps eternal death would be a better description.

> This dismal image of cold, dark, featureless near-nothingness is the closest that modern cosmology comes to the 'heat death' of nineteenth century physics. [47]

What conclusion must we draw from all this? If all life, indeed all matter, not just on earth, but throughout the universe, is doomed, then why bother about anything? The unwarranted extension of the second law beyond its actual scope of application has given rise to all manner of false and nihilistic philosophical conclusions. Thus, Bertrand Russell, the British philosopher, could write the following lines in his book *Why I Am Not a Christian*:

> All the labors of the ages, all the devotion, all the inspiration, all the noonday brightness of human genius, are destined to extinction in the vast death of the solar system, and...the whole temple of man's achievement must inevitably be buried beneath the debris of a universe in ruins — all these things, if not quite beyond dispute, are yet so nearly certain that no philosophy which rejects them can hope to stand. Only within the scaffolding of these truths, only on the firm foundation of unyielding despair, can the soul's habitation henceforth be safely built. [48]

Order Out of Chaos

In recent years, this pessimistic interpretation of the second law has been challenged by a startling new theory. The Belgian Nobel Prize winner Ilya Prigogine and his collaborators have pioneered an entirely different interpretation of the classical theories of thermodynamics. There are some parallels between Boltzmann's theories and those of Darwin. In both, a large number of random fluctuations lead to a point of irreversible change, one in the form of biological evolution, the other in that of the dissipation of energy, and evolution towards disorder. In thermodynamics, time implies degradation and death. The question arises, how does this fit in with the phenomenon of life, with its inherent tendency towards organization and ever increasing complexity.

The law states that things, if left to themselves, tend towards increased entropy. In the 1960s, Ilya Prigogine and others realized that in the real world atoms and molecules are almost never "left to themselves." Everything affects everything else. Atoms and molecules are almost always exposed to the flow of energy and material from the outside, which, if it is strong enough, can partially reverse the apparently inexorable process of disorder posited in the second law of thermodynamics. In fact, nature shows numerous instances not only of disorganization and decay, but also of the opposite processes — spontaneous self-organization and growth. Wood rots, but trees grow. According to Prigogine, self-organizing structures occur everywhere in nature. Likewise, M. Waldrop concluded:

> A laser is a self-organizing system in which particles of light, photons, can spontaneously group themselves into a single powerful beam that has every photon moving in lockstep. A hurricane is a self-organizing system powered by the steady stream of energy coming in from the sun, which drives the winds and draws rainwater from the oceans. A living cell — although much too complicated to analyze mathematically — is a self-organizing system that survives by taking in energy in the form of food and excreting energy in the form of heat and waste. [49]

Everywhere in nature we see patterns. Some are orderly, some disorderly. There is decay, but there is also growth. There is life, but there is also death. And, in fact, these conflicting tendencies are bound up together. They are inseparable. The second law asserts that all of nature is on a one-way ticket to disorder and decay. Yet this does not square with the general patterns we observe in nature. The very concept of "entropy," outside the strict limits of thermodynamics, is a problematic one.

> "Thoughtful physicists concerned with the workings of thermodynamics realize how disturbing is the question of, as one put it, "how a purposeless flow of energy can wash life and consciousness into the world." Compounding the trouble is the slippery notion of entropy, reasonably well-defined for thermodynamic purposes in terms of heat and temperature, but devilishly hard to pin down as a measure of disorder. Physicists have trouble enough measuring the degree of order in water, forming crystalline structures in the transition to ice, energy bleeding away all the while. But thermodynamic entropy fails miserably as a measure of the changing degree of form and formlessness in the creation of amino acids, of microorganisms, of self-reproducing plants and animals, of complex information systems like the brain. Certainly these evolving islands of order must obey the second law. The important laws, the creative laws, lie elsewhere. [50]

The process of nuclear fusion is an example, not of decay, but of the building-up of the universe. This was pointed out in 1931 by H. T. Poggio, who warned the prophets of thermodynamic gloom against the unwarranted attempts to extrapolate a law which applies in certain limited situations on earth to the whole universe. "Let us not be too sure that the universe is like a watch that is always running down. There may be a rewinding." [51]

The second law contains two fundamental elements — one negative and another positive. The first says that certain processes are impossible (e.g. that heat flows from a hot source to a cold one, never vice versa) and the second (which flows from the first) states that entropy is an inevitable feature of all isolated systems. In an isolated system all non-equilibrium situations produce evolution towards the same kind of equilibrium state. Traditional thermodynamics saw in entropy only a movement towards disorder. This, however, refers only to simple, isolated systems (e.g., a steam engine). Prigogine's new interpretation of Boltzmann's theories is far wider, and radically different.

Chemical reactions take place as a result of collisions between molecules. Normally, the collision does not bring about a change of state; the molecules merely exchange energy. Occasionally, however, a collision produces changes in the molecules involved (a "reactive collision"). These reactions can be speeded up by catalysts. In living organisms, these catalysts are specific proteins, called enzymes. There is every reason to believe that this process played a decisive role in the emergence of life on earth. What appear to be chaotic, merely random movements of molecules, at a certain point reach a critical stage where quantity suddenly becomes transformed into quality. And this is an essential property of all forms of matter, not only organic, but also inorganic.

> Remarkably, the perception of oriented time increases as the level of biological organization increases and probably reaches its culminating point in human consciousness. [52]

Every living organism combines order and activity. By contrast, a crystal in a state of equilibrium is structured, but inert. In nature, equilibrium is not normal but, to quote Prigogine "a rare and precarious state." Non-equilibrium is the rule. In simple isolated systems like a crystal, equilibrium can be maintained for a long time, even indefinitely. But matters change when we deal with complex processes, like living things. A living cell cannot be kept in a state of equilibrium, or it would die. The processes governing the emergence of life are not simple and linear, but dialectical, involving sudden leaps, where quantity is transformed into quality.

"Classical" chemical reactions are seen as very random processes. The molecules involved are evenly distributed in space, and their spread is distributed "normally" i.e., in a Gauss curve. These kinds of reaction fit into the concept of Boltzmann, wherein all side-chains of the reaction will fade out and the reaction will end up in a stable reaction, an immobile equilibrium. However, in recent decades chemical reactions were discovered that deviate from this ideal and simplified concept. They are known under the common name of "chemical clocks." The most famous examples are the Belousov-Zhabotinsky reaction, and the Brussels model devised by Ilya Prigogine.

Linear thermodynamics describes a stable, predictable behavior of systems that tend towards the minimum level of activity possible. However, when the thermodynamic forces acting on a system reach the point where the linear region is exceeded, stability can no longer be assumed. Turbulence arises. For a long time turbulence was regarded as a synonym for disorder or chaos. But now, it has been discovered that what appears to be merely chaotic disorder on the macroscopic (large-scale) level, is, in fact, highly organized on the microscopic (small-scale) level.

Today, the study of chemical instabilities has become common. Of special interest is the research done in Brussels under the guidance of Ilya Prigogine. The study of what happens beyond the critical point where chemical instability commences has enormous interest from the standpoint of dialectics. Of particular importance is the phenomenon of the "chemical clock." The Brussels model (nicknamed the "Brusselator" by American scientists) describes the behavior of gas molecules. Suppose there are two types of molecules, "red" and "blue," in a state of chaotic, totally random motion. One would expect that, at a given moment, there would be an irregular distribution of molecules, producing a "violet" color, with occasional flashes of red or blue. But in a chemical clock, this does not occur beyond the critical point. The system is all blue, then all red, and these changes occur at regular interval.

"Such a degree of order stemming from the activity of billions of molecules seems incredible," say Prigogine and Stengers, "and indeed, if chemical clocks had not been observed, no one would believe that such a process is possible. To change color all at once, molecules must have a way to 'communicate.' The system has to act as a whole. We will return repeatedly to this key word, communicate, which is of obvious importance in so many fields, from chemistry to neurophysiology. Dissipative structures introduce probably one of the simplest physical mechanisms for communication."

The phenomena of the "chemical clock" shows how in nature order can arise spontaneously out of chaos at a certain point. This is an important observation, especial in relation to the way in which life arises from inorganic matter.

"'Order through fluctuations' models introduce an unstable world where small causes can have large effects, but this world is not arbitrary. On the contrary, the reasons for the amplification of a small event are a legitimate matter for rational inquiry."

In classical theory, chemical reactions take place in a statistically ordered manner. Normally, there is an average concentration of molecules, with an even distribution. In reality, however, local concentrations appear which can organize themselves. This result is entirely unexpected from the standpoint of the traditional theory. These focal points of what Prigogine calls "self-organization" can consolidate themselves to the point where they affect the whole system.

What was previously thought of as marginal phenomena turn out to be absolutely decisive. The traditional view was to regard irreversible processes as a nuisance, caused by friction and other sources of heat loss in engines. But the situation has changed. Without irreversible processes, life would not be possible. The old view of irreversibility as a subjective phenomenon (a result of ignorance) is being strongly challenged. According to Prigogine irreversibility exists on all levels, both microscopic and macroscopic. For him, the second law leads to a new concept of matter. In a state of non-equilibrium, order emerges. "Non-equilibrium brings order out of chaos." [53]

9.   THE BIG BANG

Cosmology

To many people, unaccustomed to dialectical thinking, the notion of infinity is difficult to accept. It is so far at variance with the finite world of everyday objects, where everything has a beginning and an end, that it seems strange and unaccountable. Moreover, it is at variance with the teachings of most of the main world religions. Most of the ancient religions had their Creation Myth. Medieval Jewish scholars put the date of Creation at 3760 B.C., and in fact, the Jewish calendar dates from then. In 1658, Bishop Ussher worked out that the universe was created in 4004 B.C. Throughout the 18th century, the universe was considered to be six or seven thousand years old at most.

But — you might object — 20th century science has nothing in common with all these Creation myths! With modern scientific methods we can get an exact picture of the size and origins of the universe. Unfortunately, things are not as simple as that. Firstly, despite colossal advances our knowledge of the observable universe is limited by the power of even the largest telescopes, radio-signals and space-probes, to provide information. Secondly and more seriously the way in which these results and observations are interpreted in a highly speculative manner, frequently bordering on mere mysticism. All too often, one has the impression that we have indeed regressed to the world of the Creation Myth (the "Big Bang"), complete with its inseparable companion, the Day of the Final Judgement (the "Big Crunch").

Gradually, beginning with the invention of the telescope, the advance of technology has pushed the boundaries of the universe further and further away. The crystal spheres which ever since Aristotle and Ptolemy had hemmed in the minds of men were finally shattered, along with all the other barriers which the religious prejudices of the Middle Ages had placed in the way of progress.

In 1755, Kant postulated the existence of distant collections of stars, which he called "island universes." Yet as late as 1924, the entire universe was estimated to be only 200,000 light years in diameter, and consisted of just three galaxies — our own and the two neighboring ones. Then the American cosmologist, Edwin Powell Hubble, using the new 100-inch telescope at Mount Wilson, showed the Andromeda nebula to be far outside our own galaxy. Later, other galaxies were discovered still further away. Kant's "island universes" hypothesis was shown to be correct. Thus the universe was rapidly "expanded" — in the minds of men — and has continued to expand ever since, as more and more distant objects are discovered. Instead of 200,000 light years, it is now thought to measure tens of billions of light years across, and time will show that even the present calculations are nowhere near big enough. For the universe, as Nicolas of Cusa and others thought, is infinite. Before the Second World War, it was thought that the age of the universe was only two billion years. That is slightly better than Bishop Ussher's calculation. But it was still hopelessly wrong. At present there is a fierce dispute among the supporters of the big bang concerning the supposed age of the universe. We shall return to that later.

The big bang theory is really a Creation myth (just like the first book of *Genesis*). It states that the universe came into being about 15 billion years ago. Before that, according to this theory, there was no universe, no matter, no space, and, if you please, no time. At that time, all the matter in the universe is alleged to have been concentrated at a single point. This invisible dot, known to big bang aficionados as a singularity, then exploded, with such a force that it instantly filled the entire universe, which is still expanding as a result. Oh, by the way, this was the moment when "time began." In case you are wondering whether this is some kind of joke, forget it. This is precisely what the big bang theory states. This is what the great majority of university professors with long strings of letters after their names actually believe. There is the clearest evidence of a drift towards mysticism in the writings of a section of the scientific community. In recent years, we have seen a flood of books about science, which, under the guise of popular accounts of the latest theories of the universe, attempt to smuggle in religious notions of all kinds, in particular, in connection with the so-called theory of the big bang. *The New Scientist* (May 7, 1994) published an article entitled *In the Beginning Was the Bang*. The author, Colin Price, trained and worked as a scientist but is now a Congregationalist minister. He begins by asking: "Is the big bang theory disconcertingly biblical? Or to put it another way, is the *Genesis* story disconcertingly scientific?" And he ends with the confident assertion: "No one would have appreciated the big bang story more than the authors of the first two chapters of the book of *Genesis*." This is quite typical of the mystical philosophy which lies behind what Mr. Price, no doubt with tongue in cheek, but quite accurately describes as the big bang story.

The Doppler Effect

In 1915, Albert Einstein put forward his general theory of relativity. Before this, the general view of the universe was derived from the classical mechanistic model worked out in the 18th century by Sir Isaac Newton. For Newton, the universe was like a vast clockwork mechanism, obeying a number of fixed laws of motion. It was infinite in extent, but essentially unchanging. This vision of the universe suffered from the defect of all mechanistic, non-dialectical theories. It was static.

In 1929, Edwin Hubble, using a powerful new telescope, showed that the universe was far bigger than had been previously thought. Moreover, he noticed a previously unobserved phenomenon. When light reaches our eyes from a moving source, it creates a change in frequency. This may be expressed in terms of the colors of the spectrum. When a source is traveling towards us, its light is perceived to shift towards the high frequency (violet) end of the spectrum. When it moves away, we perceive a shift towards the low frequency (red) end of the spectrum. This theory, first worked out by the Austrian Christian Doppler, and called the "Doppler Effect" after him, had major implications for astronomy. The stars appear to observers as a pattern of lights against a dark background. Noticing that most of the stars showed a shift towards the red end of the spectrum, Hubble's observations gave rise to the idea that the galaxies were moving away from us at a speed proportionate to the distance of the galaxy. This became known as Hubble's Law, although Hubble himself did not think that the universe was expanding.

Hubble observed that there was a correlation between the red shift and distance, as measured by the apparent brightness of the galaxies. It appeared that the most distant galaxies then observable were moving away at 25,000 miles per second. With the advent of the new 200-inch telescope in the 1960s, even more distant objects were detected, moving away at 150,000 miles per second. Upon these observations, the hypothesis of the "expanding universe" was built. In addition, the "field equations" of Einstein's general theory of relativity could be interpreted in such a way as to make them conform to this idea. By extension, it was argued that, if the universe was expanded, it must have been smaller in the past than now. The consequence of this was the hypothesis that the universe must have begun as a single dense core of matter. This was not originally Hubble's idea. It had already been advanced in 1922 by the Russian mathematician, Alexander Friedmann. Then in 1927, George Lemaître first put forward his idea of the "cosmic egg." From the standpoint of dialectical materialism, the idea of an eternally unchanging, closed universe, in a state of permanent equilibrium, is clearly incorrect. Therefore, the abandonment of this standpoint was undoubtedly a step forward.

The theories of Friedmann were given an important boost by the observations of Hubble and Wirtz. These appeared to indicate that the universe, or at least the part of it we can observe, was expanding. This was seized upon by Georges Lemaître, a Belgian priest, who attempted to prove that, if the universe was finite in space, it must also be finite in time — it must have had a beginning. The usefulness of such a theory to the Catholic Church is beyond all doubt. It leaves the door wide open to the idea of a Creator, who, after being ignominiously expelled from the universe by science, now prepares his triumphal comeback as the Cosmic Ju-ju Man. "I felt at the time," said Hannes Alfvén years later, "that the motivation for his theory was Lemaître's need to reconcile his physics with the Church's doctrine of creation ex nihilo." [54] Lemaître was later rewarded by being made director of the Pontifical Academy of Science.

How the Theory Evolved

It is not actually correct to refer to "the big bang theory." In fact, there have been at least five different theories, each of which has run into trouble. The first, as we have seen, was put forward in 1927 by Lemaître. This was soon refuted on a number of different grounds — incorrect conclusions drawn from general relativity and thermodynamics, a false theory of cosmic rays and stellar evolution, etc. After the Second World War, the discredited theory was revived by George Gamow and others in a new form. A number of calculations were advanced by Gamow and others, (incidentally, not without a certain amount of scientific "creative accountancy") to explain the different phenomena which would flow from the big bang — density of matter, temperature, radiation levels, and so on. George Gamow's brilliant style of writing ensured that the big bang captured the popular imagination. Once again, the theory ran up against serious problems.

A whole number of discrepancies were found which invalidated, not only Gamow's model, but the "oscillating universe" model subsequently worked out by Robert Dicke and others, in an attempt to get round the problem of what happened before the big bang, by making the universe oscillate in a never-ending cycle. But Gamow had made one important prediction — that such an immense explosion would leave behind evidence in the form of "background radiation," a kind of echo of the big bang in space. This was used to revive the theory some years later.

From the beginning there was opposition to the idea. In 1928 Thomas Gold and Hermann Bondi advanced the "steady state" as an alternative, later popularized by Fred Hoyle. While accepting the expanding universe, it attempted to explain it by the "continuous creation of matter from nothing." This was alleged to be happening all the time, but at a rate too slow to be detected by present-day technology. This means that the universe remains essentially the same for all time, hence the "steady state" theory. Thus matters went from bad to worse. From the "cosmic egg" to matter created out of nothing! The two rival theories slugged it out for over a decade.

The very fact that so many serious scientists were prepared to accept Hoyle's fantastic notion that matter was being created out of nothing is itself absolutely astonishing. In the event, this theory was shown to be false. The steady state theory assumed the universe to be homogeneous in time and space. If the universe were in a "steady state" for all time, the density of a radio-emitting object ought to be constant, since the further we look out into space, the further back in time we see. However, observations showed that this was not the case; the further they looked out into space, the greater the intensity of the radio waves. This proved conclusively that the universe was in a constant state of change and evolution. It had not always been the same. The steady state theory was wrong.

In 1964, the steady state theory received the coup de grace with the discovery by two young astronomers in the USA, Arnas Penzias and Robert Wilson, of background radiation in space. This was immediately taken to be the "after-echo" of the big bang, predicted by Gamow. Even so, there were inconsistencies. The temperature of the radiation was found to be only 3.5°K, not the 20°K predicted by Gamow, or the 30°K predicted by his successor, P. J. E. Peebles. This result is even worse than it looks. Since the amount of energy in a field is proportional to the 4th power of its temperature, the energy of the observed radiation was actually several thousand times less than that predicted.

Robert Dicke and P. J. E. Peebles took over the theory where Gamow had left off. Dicke realized that there was a handy way of getting round the sticky question of what happened before the big bang, if only they could get back to Einstein's idea of a closed universe. It could then be argued that the universe would expand for a time, then collapse to a single point (a "singularity"), or something near it, and then bounce back into expansion, in a kind of everlasting cosmic ping-pong game. The trouble was that Gamow had calculated the energy and density of the universe at levels just short of what would be needed to close the universe. The density was about two atoms per cubic meter of space; and the energy density, expressed as the predicted temperature of the background radiation, supposed to represent the remnants of the big bang, 20°K, i.e., 20 degrees above absolute zero. In fact, Gamow had fixed these figures in order to prove that the big bang produced heavy elements, something nobody now accepted. So Dicke unceremoniously ditched them, and selected new and equally arbitrary figures, which would fit in with his theory of a closed universe.

Dicke and Peebles predicted that the universe would be filled with radiation, mainly radio waves, with a temperature of 30°K. Later, Dicke claimed his group had predicted a temperature of 10°K, although this figure does not appear anywhere in his published notes, and is anyway a 100 times more than the observed result. This showed that the universe was more diffuse than Gamow had thought, with less gravity, which aggravated the basic problem of where all the energy for the big bang came from.

As Eric Lerner points out: "Far from confirming the Peebles-Dicke model, the Penzias-Wilson discovery clearly ruled out the closed oscillating model." [55] Thus arose a third version of the big bang — which became known as the standard model — an open universe in a permanent state of expansion.

Fred Hoyle did some detailed calculations, and announced that a big bang would produce only light elements — helium, deuterium and lithium (the latter two are actually quite rare). He calculated that if the density of the universe were about one atom per eight cubic meters, the amounts of these three light elements would be quite close to those actually observed. In this way, a new version of the theory was put forward which was nothing like the older theories. This no longer mentioned the cosmic rays of Lemaître, or the heavy elements of Gamow. Instead, the evidence put forward was the microwave background and three light elements. Yet none of this constitutes conclusive proof for the big bang. A major problem was the extreme smoothness of the background microwave radiation. The so-called irregularities in the background are so small that these fluctuations would not have had time to grow into galaxies — not unless there was a lot more matter (and therefore a lot more gravity) around than appears to be the case.

There were other problems, too. How does it come about that bits of matter flying in opposite directions all managed to reach the same temperature, and all at the same time (the "horizon" problem)? The partisans of the theory present the alleged origins of the universe as a model of mathematical perfection, all perfectly regular, a regular "Eden of symmetry whose characteristics conform to pure reason," as Lerner puts it. But the present universe is anything but perfectly symmetrical. It is irregular, contradictory, "lumpy." Not at all the stuff that well-mannered equations are made of down at Cambridge! One of the problems is why did the big bang not produce a smooth universe? Why did not the original simple material and energy just spread out evenly in space as an immense haze of dust and gas? Why is the present universe so "lumpy"? Where did all these galaxies and stars come from? So how did we get from A to B? How did the pure symmetry of the early universe give rise to the present irregular one we see before our eyes?

The "Inflation" Theory

To get round this and other problems, Alan Guth, the American physicist, advanced his theory of the "inflationary universe." (It may be no coincidence that this idea was put forward in the 1970s, when the capitalist world was going through an inflationary crisis!) According to this theory, the temperature dropped so rapidly that there was no time for the different fields to separate out or for different particles to form. The differentiation took place only later, when the universe was much larger. This, then, is the most recent version of the big bang. It asserts that, at the time of the big bang, the universe experienced an exponential expansion, in which it doubled in size every 10–35 seconds (hence "inflation"). Whereas the earlier versions of the "standard model" envisaged the whole of the universe squashed to the size of a grapefruit, Guth went one better. He calculated that the universe did not begin as a grapefruit, but instead, it would be a billion times smaller than the nucleus of a hydrogen atom. Then it would expand at an incredible speed — many times the speed of light, which is 186,000 miles per second — until it reached a size 1090 times its initial volume, that is, 1 with 90 zeros after it!

Let us examine the implications of this theory. Like all the other big bang theories, it sets out from the hypothesis that all the matter in the universe was concentrated in a single spot. The fundamental mistake here is to imagine that the universe is equal to the observable universe, and that it is possible to reconstruct the entire history of the universe, as a linear process, without taking into account all the different phases, transitions, and different states through which matter passes.

Dialectical materialism conceives of the universe as infinite, but not static or in a permanent state of "equilibrium," as both Einstein and Newton did. Matter and energy cannot be created or destroyed, but are in a continual process of movement and change, which involves periodic explosions, expansion and contraction, attraction and repulsion, life and death. There is nothing intrinsically improbable about the idea of one, or many, great explosions. The problem here is a different one — a mystical interpretation of certain observed phenomena, such as the Hubble red shift, and an attempt to smuggle the religious idea of the creation of the universe into science by the back door.

To begin with, it is unthinkable that all the matter in the universe should be concentrated in a single point "of infinite density." Let us be clear what this means. Firstly, it is impossible to place an infinite amount of matter and energy in a finite space. Just to pose the question is sufficient to answer it. "Ah! say the

Big Bangers, but the universe is not infinite, but finite, according to Einstein's general theory of relativity." In his book, Eric Lerner points out that an infinite number of different universes are allowed by Einstein's equations. Friedmann and Lemaître showed that many equations led to universal expansion. But by no means all of them imply a state of "singularity." Yet this is the one variant which is dogmatically advanced by Guth and co.

Even if we accept that the universe is finite, the notion of "singularity" leads us to conclusions of a clearly fantastic character. If we take the tiny corner of the universe which we are able to see as being the whole universe — an arbitrary assumption with no logical or scientific basis whatsoever — then we are talking about more than 100 billion galaxies, each containing about 100 billion main sequence stars (like our own sun). According to Guth, all this matter was concentrated in a space smaller than a single proton. When it had existed for a millionth of a trillionth of a trillionth, of a trillionth of a second with a temperature of trillions of trillions of trillions of degrees, there was only one field and only one kind of particle interaction. As the universe expanded and the temperature fell, the different fields are supposed to have "condensed" out of the original state of simplicity.

The question arises where all the energy came from to propel such an unprecedented expansion. In order to solve this riddle, Guth resorted to a hypothetical omnipresent force field (a "Higgs field"), the existence of which is predicted by some theoretical physicists, but for which there is not a shred of empirical evidence. "In Guth's theory," comments Eric Lerner, "the Higgs field which exists in a vacuum generates all the needed energy from nothing — ex nihilo. The universe, as he puts it, is one big 'free lunch,' courtesy of the Higgs field." [56]

Dark Matter?

Every time the big bang hypothesis runs into trouble, instead of abandoning it, its supporters just move the goal posts, introducing new and ever more arbitrary assumptions in order to shore it up. For example, the theory requires a certain amount of matter in the universe. If the universe was created 15 billion years ago, as the model predicts, there has simply not been enough time for the matter we observe to have congealed into galaxies like the Milky Way, without the help of invisible "dark matter." According to the big bang cosmologists, in order for galaxies to have been formed from the big bang, there must have been sufficient matter in the universe to bring about an eventual halt to its expansion through the law of gravitation. This would mean a density of approximately ten atoms per cubic meter of space. In reality, the amount of matter present in the observable universe is about one atom per ten cubic meters — a hundred times less than the amount predicted by the theory.

The cosmologists decided to represent the density of the universe as a ratio of the density needed to bring the expansion to a halt. They call this ratio omega. Thus, if omega equals 1, it would just be sufficient to halt the expansion. Unfortunately, the actual ratio was observed to be around .01 or .02. Approximately 99% of the required matter had somehow "gone missing." How to solve the conundrum? Very simply. Since the theory demanded that the matter be there, they arbitrarily fixed the value of omega at close to 1, and then began a frantic search for the missing matter! The first problem facing the big bang was the origin of the galaxies. How did the extremely smooth background radiation produce such a "lumpy" irregular universe? The so-called "ripples" (anisotropies) in the radiation were supposed to have been a reflection of the formation of the clumps of matter around which the early galaxies coalesced. But the irregularities observed were too small to have been responsible for the formation of galaxies, unless there was a lot more matter, and therefore gravity, present than seems to be the case. To be exact, there needed to be 99% more matter, which just wasn't there.

This is where the notion of "cold dark matter" comes in. It is important to realize that no-one has ever seen this stuff. Its existence was put forward just over ten years ago, in order to fill up an embarrassing hole in the theory. Since only 1 or 2% of the universe can actually be seen, the remaining 99% or so was alleged to consist of invisible matter, which is dark and cold, emitting no radiation at all. Such strange particles, after a decade of searching for them, remain unobserved. But they nevertheless occupy a central place in the theory, simply because it demands that they should exist.

Fortunately, it is possible to work out quite accurately the amount of matter in the observable universe. It is about one atom for every ten cubic meter of space. This is a hundred times less than the amount required by the big bang theory. But, as the journalists like to say, don't let the facts spoil a good story! If there is not enough matter in the universe to square with the theory, then there must be an awful lot of matter there which we can't see. As Brent Tully put it, "It's disturbing to see that there is a new theory every time there's a new observation."

At this stage, the defenders of the big bang decided to call on the aid of the Seventh Cavalry, in the person of particle physicists. The mission they were called upon to carry out puts all the exploits of John Wayne completely in the shade. The most he ever had to do was to find some unfortunate women and children carried off by the Indians. But when the cosmologists called in their colleagues who were busy investigating the mysteries of "inner space," their request was a trifle more ambitious. They wanted them to find the 99% or so of the universe which had inconsiderately "gone missing." Unless they could find this missing matter, their equations would just not add up, and the standard theory of the origin of the universe would be in trouble!

In his book *The Big Bang Never Happened* Eric Lerner details a whole series of observations, the results of which have been published in scientific journals, which completely refute the idea of dark matter. Yet, in the teeth of all the evidence, the advocates of the big bang continue to behave like the learned professor who refused to look through the telescope to test the correctness of Galileo's theories. Dark matter must exist — because our theory demands it!

> The test of scientific theory [writes Lerner], is the correspondence of predictions and observation, and the big bang has flunked. It predicts that there should be no objects in the universe older than twenty billion years and larger than 150 million light-years across. There are. It predicts that the universe, on such a large scale, should be smooth and homogeneous. The universe isn't. The theory predicts that, to produce the galaxies we see around us from the tiny fluctuations evident in the microwave background, there must be a hundred times as much dark matter as visible matter. There's no evidence that there's any dark matter at all. And if there is no dark matter, the theory predicts, no galaxies will form. Yet there they are, scattered across the sky. We live in one. [57]

Alan Guth succeeded in removing some of the objections to the big bang, but only by advancing the most fantastic and arbitrary version of the theory yet seen. It did not say what the "dark matter" was, but merely provided the cosmologists with a theoretical justification for it. The real significance was that it established the link between cosmology and particle physics which has lasted ever since. The problem is that the general tendency of theoretical physics, as in cosmology, has been to resort increasingly to a priori mathematical assumptions to justify their theories, making very few predictions that can be tested in practice. The resulting theories have an ever more arbitrary and fantastic character, and frequently seem to have more in common with science fiction than anything else.

In point of fact, the particle physicists who rushed to the aid of cosmology had plenty of problems of their own. Alan Guth and others were trying to discover a Grand Universal Theory (GUT), which would unify the three basic forces which operate on the small scale in nature — electromagnetism, the weak force (which causes radioactive decay), and the strong force (which holds the nucleus together, and is responsible for the release of nuclear energy). They hoped to repeat the success of Maxwell, a hundred years earlier, who had proved that electricity and magnetism were one and the same force. The particle physicists were only too willing to enter an alliance with the cosmologists, in the hope of finding the answer in the heavens for the difficulties they had found themselves in. In reality, their whole approach was similar. With scarcely any reference to observation, they based themselves on a series of mathematical models, and completely arbitrary assumptions, which were often little more than mere speculation. Theories have emerged thick and fast, each more incredible than the last. "Inflation" theory is mixed up with all this.

The Neutrino to the Rescue!

The determination with which the supporters of the big bang cling to their positions frequently leads them to perform the most amusing somersaults. Having searched in vain for the 99% of missing "cold dark matter," they failed to find anything like the quantities required by the theory, to prevent the universe from expanding forever. On 18th December 1993, *The New Scientist* published an article entitled *Universe Will Expand Forever*. Here it was admitted that "a group of galaxies in the constellation of Cepheus contains far less invisible matter than had been thought a few months ago," and that the claims made earlier by American astronomers was "based on faulty analysis." A lot of scientific reputations are at stake, not to mention hundreds of millions of dollars in research grants. Could this fact have some connection with the fanaticism with which the big bang is defended? As usual, they saw what they wanted to see. The facts had to conform with the theory!

The evident failure to find the "cold dark matter," the existence of which is essential to the survival of the theory, was causing unease in the more thinking sections of the scientific community. An editorial of *The New Scientist*, published on the 4th June 1994 with the suggestive title *A Folly of Our Time?* compared the idea of dark matter with the discredited Victorian concept of the 'ether,' an invisible medium, by which light waves were believed to travel through space:

> It was invisible, ubiquitous, and, in the late 19th century, every physicist believed in it. It was, of course, the aether, the medium in which they thought light propagated, and it turned out to be a phantom. Light does not need a medium in which to propagate, unlike sound.

> Now, at the close of the 20th century, physicists find themselves in a curiously similar situation to their Victorian counterparts. Once again they are putting their faith in something which is invisible and ubiquitous. This time it is dark matter.

At this point, one would expect a serious scientist to begin to ask himself whether there was not something basically wrong with his theory. The same editorial adds:

> In cosmology, free parameters seem to be proliferating like wildfire. If the observations do not fit the theory, cosmologists seem happy to simply add new variables. By continually patching up the theory, we may be missing out on some Big Idea.

Indeed. But, don't let the "facts" get in the way. Like a conjurer pulling a rabbit out of a hat, they suddenly discovered — the neutrino!

The neutrino, which is a subatomic particle, is described by Hoffmann as "fluctuating uncertainly between existence and non-existence." That is to say, in the language of dialectics, "it is and is not." How can such a phenomenon be reconciled with the law of identity, which categorically asserts that a thing either is or is not? Faced with such dilemmas, which reappear at every step in the world of subatomic particles described by quantum mechanics, there is frequently a tendency to resort to formulations such as the idea that the neutrino is a particle with neither mass nor charge. The initial opinion, still held by many scientists, was that the neutrino had no mass, and since electric charge cannot exist without mass, the inescapable conclusion was that the neutrino had neither.

Neutrinos are extremely small particles, and therefore difficult to detect. The existence of the neutrino was first postulated to explain a discrepancy in the amount of energy present in particles emitted from the nucleus. A certain amount of energy appeared to be lost, which could not be accounted for. Since the law of the conservation of energy states that energy can neither be created nor destroyed, this phenomenon required another explanation. Although it seems that the idealist physicist Niels Bohr was quite prepared to throw the law of conservation of energy overboard in 1930, this proved to be slightly premature! The discrepancy was explained by the discovery of a previously unknown particle — the neutrino.

Neutrinos formed in the sun's core at a temperature of 15 million degrees centigrade moving at the speed of light reach the sun's surface in three seconds. Floods of them stream through the universe, passing through solid matter, apparently without interacting on it. Neutrinos are so small that they pass straight through the earth. So tiny are these elusive particles that their interaction with other forms of matter is minimal. They can pass through the earth, and even through solid lead, leaving no trace. Indeed, trillions of neutrinos are passing through your body even as you read these lines. But the likelihood that one could be trapped there is negligible, so you needn't worry. It has been estimated that a neutrino can pass through solid lead with a thickness of 100 light-years, with only a 50 per cent chance of being absorbed. That is why it remained undetected for so long. Indeed, it is difficult to imagine how a particle, which is so small that it was thought to have neither mass nor charge, and can pass through 100 light-years of lead, could ever be detected. But detected it was.

It seems that some neutrinos can be stopped by the equivalent of one tenth of an inch of lead. In 1956, using an ingenious experiment, American scientists succeeded in trapping an anti-neutrino. Then in 1968, they discovered neutrinos from the sun, although only one-third of the amount predicted by the current theories. Undoubtedly the neutrino possessed properties which could not immediately be detected. Given its extreme smallness, that was not surprising. But the idea of a form of matter which lacked the most basic properties of matter was clearly a contradiction in terms. In the event, the problem appears to have been resolved from two completely different sources. First, one of the discoverers of the neutrino, Frederick Reines, announced in 1980 that he had discovered the existence of neutrino oscillation in an experiment. This would indicate that the neutrino does have mass, but Reines' results were not seen as conclusive.

However, Soviet physicists, involved in an entirely separate experiment, showed that electron-neutrinos have a mass, which could be as much as 40 electron volts. Since this is only 1/13,000th of the mass of an electron, which in turn is only 1/2000th of a proton, it is hardly surprising that the neutrino was for so long believed to have no mass.

Up till recently, the general view of the scientific establishment was that the neutrino had no mass and no charge. Now, all of a sudden, they have changed their mind and declared that the neutrino does indeed have mass — and, perhaps, quite a lot of it. This is the most astonishing conversion since Saint Paul fell off his horse on the road to Damascus! Indeed, such indecent haste must raise serious doubts about the motivation behind this miraculous conversion. Can it be that they were so desperate at their signal failure to deliver the goods with "cold dark matter" that they finally decided to do an about-turn on the neutrino? One can just imagine what Sherlock Holmes would have said to Doctor Watson!

Despite the enormous advances in the field of particle research, the present situation is confused. Hundreds of new particles have been discovered, but as yet there is no satisfactory general theory capable of introducing some order, as Mendeleyev did in the field of chemistry. At present, there is an attempt to unify the fundamental forces of nature by grouping them under four headings: gravity, electromagnetism, and the "weak" and "strong" nuclear forces, each of which functions at a different level.

Gravitation works on the cosmological scale, holding the stars, planets and galaxies together. Electromagnetism binds atoms into molecules, transports photons from the sun and stars, and fires the synapses of the brain. The strong force binds together protons and neutrons inside the nuclei of atoms. The weak force is expressed in the transmutation of unstable atoms during radioactive decay. Both the latter forces only operate at very short range. However, there is no reason to suppose that this arrangement represents the last word on the subject, in some respects it is an arbitrary notion.

There are big differences between these forces. Gravitation affects all forms of matter and energy, whereas the strong force only affects one class of particles. Yet gravitation is one hundred million trillion trillion trillion times weaker than the strong nuclear force. More importantly, it is not evident why there should be no opposite force to gravity, whereas electromagnetism is manifested both as positive and negative electrical charge. This problem, the solution of which was attempted by Einstein, remains to be solved, and has a vital bearing on the entire discussion about the nature of the universe. Each force is accounted for by a different set of equations, involving some twenty different parameters. These give results, but nobody knows why.

The so-called Grand Unified Theories ("GUTs") put forward the idea that matter itself might only be a passing phase in the evolution of the universe. However, the prediction made by the GUTs that protons decay has not been borne out, thus invalidating at least the simplest version of the GUTs. In an attempt to make sense of their own discoveries, some physicists have got entangled in ever more weird and wonderful theories, like the so-called "supersymmetry" theories ("SUSYs") which purport that the universe was originally built on more than four dimensions. According to this notion, the universe could have started with, for example, ten dimensions, but unfortunately all but four of them collapsed during the big bang, and are now too small to be noticed.

Apparently, these objects are the subatomic particles themselves, which are alleged to be quanta of matter and energy that condensed out of pure space. Thus they stagger from one metaphysical speculation to the next in a vain attempt to explain the fundamental phenomena of the universe. Supersymmetry postulates the universe as beginning in a state of absolute perfection. In the words of Stephen Hawking, "the early universe was simpler, and it was a lot more appealing, because it was a lot simpler." Some scientists even try to justify this kind of mystical speculation on aesthetic grounds. Absolute symmetry is alleged to be beautiful. Thus we find ourselves back in the rarefied atmosphere of Plato's idealism.

In reality, nature is not characterized by absolute symmetry, but is full of contradictions, irregularities, cataclysms, and sudden breaks of continuity. Life itself is a proof of this assertion. In any living system, absolute equilibrium signifies death. The contradiction which we observe here is as old as the history of human thought. It is the contradiction between the "perfect" abstractions of thought and the necessary irregularities and "imperfections" which characterize the real material world. The whole problem stems from the fact that the abstract formulae of mathematics, which may or may not be beautiful, most certainly do not adequately represent the real world of nature. To suppose such a thing is a methodological error of the first magnitude, and necessarily leads us to draw wrong conclusions.

Constant Headaches, or Hubble Trouble

At present there is a fierce dispute among the supporters of the big bang concerning the supposed age of the universe. In fact, the entire "standard model" is in crisis. We are treated to the spectacle of respectable people of science attacking each other in public with the most ungentlemanly language. And all over something called the Hubble Constant. This is the formula which measures the speed at which things are moving in the universe. This is of great importance for those who wish to discover the age and size of the universe. The trouble is that nobody knows what it is!

Edwin Hubble asserted that the speed with which the galaxies are moving apart was proportional to their distance from us — the further away, the faster they are moving. This expressed in Hubble's Law — v(elocity) = H x d(istance). In this equation, the H is known as Hubble's Constant. In order to measure this, we need to know two things: the speed and distance away of a particular galaxy. The speed can be calculated by the red shift. But the distance between galaxies cannot be measured with a slide-rule. In fact, no reliable instruments exist for measuring such immense distances. And here lies the rub! The experts cannot agree on the real value of the Hubble Constant, as was comically revealed in a recent Channel Four (UK) T.V. program:

"Michael Pierce says that, without doubt, the Hubble Constant is 85, Gustaf Tamman asserts 50, George Jacoby 80, Brian Schmidt 70, Michael Rowan Robinson 50, and John Tonry 80. The difference between 50 and 80 may not sound like much," says the accompanying Channel Four booklet, "but it is crucial to the age of the Universe. If the Hubble is high, astronomers could be in the process of disproving their most important theory."

The importance of this is that the higher the "Hubble," the faster things are moving, and the sooner in the past was the moment when the big bang was supposed to have occurred. In recent years, new techniques of measuring the distance of galaxies have been applied, which have led astronomers to revise earlier estimates drastically. This has provoked consternation in the scientific community, since the estimates for the Hubble Constant have been getting higher all the time. The latest estimate puts the age of the universe at just 8 billion years. This would mean that there are stars which are older than the universe itself! This is a glaring contradiction - not a dialectical one, but simply nonsense.

"Well," comments Carlos Frank, quoted in the same booklet, "if it turns out that the ages of the stars are greater than the expansion time of the universe, as inferred by the measurement of the Hubble Constant and the measurement of the density of the universe, then there is a genuine crisis. You only have one option: you have to drop the basic assumptions upon which the model of the Universe is based. In this case, you have to drop some, perhaps all, of the basic assumptions on which the big bang theory is based." [58]

There is virtually no empirical evidence to bear out the big bang theory. Most of the work done to support it is of a purely theoretical character, leaning heavily on abstruse and esoteric mathematical formulae. The numerous contradictions between the preconceived "big bang" schema and the observable evidence have been covered up by constantly moving the goal posts in order to preserve at all costs a theory upon which so many academic reputations have been built.

According to this theory, there can be nothing in the universe older than 15 billion years. But there is evidence that contradicts this proposition. In 1986, Brent Tully of Hawaii University discovered huge agglomerations of galaxies ("superclusters") about a billion light years long, three hundred million light years wide and one hundred million light-years thick. In order for such vast objects to form, it would have taken between eighty and a hundred billion years, that is to say four or five times longer than what would be allowed by the "big bangers." Since then there have been other results which tend to confirm these observations.

The New Scientist (5th February, 1994) carried a report of the discovery of a cluster of galaxies by Charles Steidel of the Massachusetts Institute of Technology and Donald Hamilton of the California Institute of Technology in Pasadena with big implications for the big bang theory:

"The discovery of such a cluster spells trouble for theories of cold dark matter, which assume that a large fraction of the mass of the universe is in cold, dark objects such as planets or black holes. The theories predict that material in the early universe clumped together from the 'bottom up,' so that galaxies formed first, then only later clumped to form clusters."

As usual, the initial reaction of astronomers is to resort to "move the goal posts," adjusting the theory to get round awkward facts. Mauro Giavalisco of the Baltimore Space Telescope Science Institute "believes it might just be possible to explain the birth of the first galaxy cluster at a red shift of 3.4 by fine-tuning the cold dark matter theory. But he adds a warning. 'If you found ten clusters at red shift 3.5, it would kill cold dark matter theories.'" [

We may take for granted that not just ten but a far larger number of these vast clusters exist and will be discovered. And these, in turn, will only represent a minute proportion of all the matter which stretches far beyond the limits of the observable universe and reaches out to infinity. All attempts to place a limit on the material universe are doomed to fail. Matter is boundless, both at the subatomic level, and with regard to time and space.

Big Crunch and Superbrain

> *Dies irae, dies illa*
> *Solvet saeclum in favilla.*
> (Thomas of Celano, *Dies Irae*)

> That day, the day of wrath,
> will turn the universe to ashes.
> — Mediaeval Church chant for the dead.

In the same way they cannot agree on the origin of the universe, so they also disagree on how it is all supposed to end up — except that they all agree that it will end badly! According to one school of thought, the expanding universe will eventually be brought to a halt by the force of gravity, whereupon the whole thing will collapse in on itself, leading to a "big crunch," where we will all end up just where we started, back inside the cosmic egg. Not so! exclaims another school of big bangers. Gravity is not strong enough to do this. The universe will simply keep on expanding indefinitely, getting thinner and thinner, like "Augustus who would not have any soup," until eventually it fades away into the black night of nothingness.

Decades ago, Ted Grant, using the method of dialectical materialism, showed the unsoundness both of the big bang theory of the origin of the universe and the alternative steady state theory put forward by Fred Hoyle and H. Bondi. Subsequently, the steady state theory, which was based on the continuous creation of matter (from nothing) was shown to be false. The big bang theory therefore "won" by default, and is still defended by the majority of the scientific establishment. From the standpoint of dialectical materialism, it is arrant nonsense to talk about the "beginning of time," or the "creation of matter." Time, space, and motion are the mode of existence of matter, which can neither be created nor destroyed. The universe has existed for all time, as constantly changing, moving, evolving matter and (which is the same thing) energy. All attempts to find a "beginning" or an "end" to the material universe will inevitably fail. But how is one to explain this strange regression to a mediaeval view of the fate of the universe?

While it is pointless to look for a direct causal link between the processes at work in society, politics and the economy, and the development of science (the relationship is neither automatic nor direct, but far more subtle), it is hard to resist the conclusion that the pessimistic outlook of some scientists in relation to the future of the universe is not accidental, but somehow related to a general feeling that society has reached an impasse. The end of the world is nigh. This is not a new phenomenon. The same doom-laden outlook was present in the period of decline of the Roman Empire and at the close of the Middle Ages. In each case, the idea that the world was coming to an end reflected the fact that a particular system of society had become exhausted and was on the point of extinction. What was imminent was not the end of the world, but the collapse of slavery and feudalism.

Just take the following quote from *The First Three Minutes* by Nobel Prize winner Steven Weinberg:

> It is almost irresistible for humans to believe that we have some special relation to the universe, that human life is not just a more or less farcical outcome of a chain of accidents reaching back to the first three minutes, but that we were somehow built in from the beginning. As I write this I happen to be in an airplane at 30,000 feet, flying over Wyoming en route home from San Francisco to Boston. Below, the earth looks very soft and comfortable — fluffy clouds here and there, snow turning pink as the sun sets, roads stretching straight across the country from one town to another. It is very hard to realize that this all is just a tiny part of an overwhelmingly hostile universe. It is even harder to realize that this present universe has evolved from an unspeakably unfamiliar early condition, and faces a future extinction of endless cold or intolerable heat. The more the universe seems comprehensible, the more it also seems pointless. [59]

We have already seen how the big bang theory opens the door to religion and all kinds of mystical ideas. To blur the distinction between science and mysticism is to put back the clock 400 years. It is a reflection of the current irrational mood of society. And it invariably leads to conclusions of a thoroughly reactionary nature. Let us take just one apparently remote and obscure question: "Do protons decay?" As we have said, this is one of the predictions of one of the branches of modern particle physics known as the GUTs. All kinds of sophisticated experiments were conducted to test this. All ended in complete failure. Yet they persist in putting forward the same idea.

Here is a typical example of the type of literature issued by the advocates of the big crunch theory:

> In the final moments, gravity becomes the all-dominant force, mercilessly crushing matter and space. The curvature of space-time increases ever faster. Larger and larger regions of space are compressed into smaller and smaller volumes. According to conventional theory, the implosion becomes infinitely powerful, crushing all

matter out of existence and obliterating every physical thing, including space and time themselves, at a space-time singularity.

This is the End.

The "big crunch," as far as we understand it, is not just the end of matter. It is the end of everything. Because time itself ceases at the big crunch, it is meaningless to ask what happens next, just as it is meaningless to ask what happened before the big bang. There is no "next" for anything at all to happen — no time even for inactivity nor space for emptiness. A universe that came from nothing in the big bang will disappear into nothing at the big crunch, its glorious few zillion years of existence not even a memory.

The question that follows is a classic of unconscious humor:

Should we be depressed by such a prospect?

— Paul Davies asks, presumably expecting a serious answer! He then proceeds to cheer us up by speculating on various means whereby humankind might escape destruction. Inevitably, we immediately find ourselves in a kind of never-never land half way between religion and science fiction.

One might wonder whether a superbeing inhabiting the collapsing universe in its final moments could have an infinite number of distinct thoughts and experiences in the finite time available. [So, before the final three minutes are up, humanity casts off its crude material body, and becomes pure spirit, able to survive the ending of everything by transforming itself into a Superbrain.]

Any superbrain would need to be quick-witted and switch communications from one direction to another as the oscillations brought more rapid collapse in one direction and then another. If the being can keep pace, the oscillations could themselves provide the necessary energy to drive the thought processes. Furthermore, in simple mathematical models there appears to be an infinite number of oscillations in the finite duration terminating in the big crunch. This provides for an infinite amount of information processing, hence, by hypothesis, an infinite subjective time for the superbeing. Thus the mental world may never end, even though the physical world comes to an abrupt cessation at the big crunch. [60]

One really needs a Superbrain to make head or tail of this! It would be nice to think that the author is joking. Unfortunately, we have read too many passages of this kind recently to be sure of this. If the Big Crunch signifies "the end of everything," where does this leave our friend the Superbrain? To begin with, only an incorrigible idealist could conceive of a brain without a body. Of course, we are here in the presence, not of any old brain, but a Superbrain. But even so, we assume that the presence of a spinal cord and a central nervous system would be of some use to it; that such a nervous system ought in all fairness, to posses a body; and that a body (even a Superbody) generally requires some kind of sustenance, specially since the brain is known to be somewhat greedy, and absorbs a very high percentage of the total calories consumed even by a mere mortal. A Superbrain would logically possesses a Superappetite! Sadly, since the big crunch is the end of everything, our unfortunate Superbrain will evidently be placed in a rather strict diet for the rest of eternity. We can only hope that, being quick-witted, it will have had time to snatch a quick meal before its three minutes was up. With this edifying thought, we take our leave of the Superbrain, and return to reality.

Is it not astonishing that, after two thousand years of the greatest advances of human culture and science, we find ourselves back in the world of the Book of Revelations? Engels warned a hundred years ago that, by turning their backs on philosophy, scientists would inevitably end up in the "spirit world." Unfortunately, his prediction has proven to be all too accurate.

A "Plasma Universe"?

The standard model of the universe has led us into a scientific, philosophical, and moral dead-end. The theory itself is full of holes. Yet it still remains on its feet, though badly shaken, mainly for the lack of an alternative. Nevertheless, something is stirring in the world of science. New ideas are beginning to take shape, which not only reject the big bang, but which set out from the idea of an infinite, constantly changing universe. It is far too early to say which of these theories will be vindicated. One interesting hypothesis, that of the "plasma universe," has been put forward by the Swedish Nobel Prize winning physicist Hannes Alfvén. While we cannot deal with the theory in detail, we feel we should at least mention some of Alfvén's ideas.

Alfvén passed from the investigation of plasma in the laboratory to a study of how the universe evolves. Plasma consists of hot, electrically conducting gases. It is now known that 99% of the matter in the universe is plasma. Whereas in normal gases, electrons are bound to an atom and cannot move easily, in a plasma, the electrons are stripped off by intense heat, allowing them to move freely. Plasma cosmologists envisage a universe "crisscrossed by vast electrical currents and powerful magnetic fields, ordered by the cosmic counterpoint of electromagnetism and gravity." [61] In the 1970s, the Pioneer and Voyager spacecraft detected the presence of electrical currents and magnetic fields filled with plasma filaments around Jupiter, Saturn and Uranus.

Scientists like Alfvén, Anthony Peratt and others, have elaborated a model of the universe which is dynamic, not static, but which does not require a beginning in time. The phenomenon of the Hubble expansion needs an explanation. But the big bang is not necessarily it. A big bang will certainly produce an expansion, but an expansion does not require a big bang. As Alfvén puts it: "This is like saying that because all dogs are animals, all animals are dogs." The problem is not the idea of an explosion, which at some point gave rise to an expansion of part of the universe. There is nothing intrinsically improbable in this. The problem is the idea that all the matter in the universe was concentrated at a single point, and that the universe and time itself was born in a single instant called the big bang.

The alternative model suggested by Hannes Alfvén and Oskar Klein accepts that there could have been an explosion, caused by the combination of large amounts of matter and antimatter in one small corner of the visible universe, which generated huge qualities of energetic electrons and positrons. Trapped in magnetic fields, these particles drove the plasma apart for hundreds of millions of years. "The explosion of this epoch, some ten or twenty billion years ago, sent the plasma from which the galaxies then condensed flying outward — in the Hubble expansion. But this was in no way a big bang that created matter, space, and time. It was just a big bang, an explosion in one part of the universe. Alfvén is the first to admit that this explanation is not the only possible one. 'The significant point,' he stresses, 'is that there are alternatives to the big bang.'"

At a time when almost all other scientists believed that space was an empty vacuum, Alfvén showed that this was not the case. Alfvén pointed out that the entire universe is pervaded by plasma currents and magnetic fields. Alfvén did pioneering work in the field of sunspots and magnetic fields. Later, Alfvén proved that when a current flows through a plasma in the laboratory, it assumes the form of a filament in order to move along magnetic field lines. Starting out from this observation, he then concluded that the same phenomenon takes place in plasma in space. It is a general property of plasma throughout the universe. Thus, we have immense electrical currents flowing along naturally-formed plasma filaments, which crisscross the cosmos.

> By forming the filamentary structures observed on the smallest and largest scales, matter and energy can be compressed in space. But it is clear that energy can be compressed in time as well — the universe is filled with sudden, explosive releases of energy. One example that Alfvén was familiar with is the solar flare, the sudden release of energy on the sun's surface, which generates the streams of particles that produce magnetic storms on earth. His 'generator' models of cosmic phenomena showed how energy can be produced gradually, as in a well-behaved power station, but not explosively, as in the flares. Understanding the explosive release of energy was the key to the dynamics of the cosmos.

Alfvén had proved the correctness of the Kant-Laplace Nebular Hypothesis. Now, if the stars and planets can be formed by the action of huge filamentary currents, there is no reason why whole solar systems cannot be formed in the same way:

> Again, the process is identical, but this time immensely larger: filaments sweeping through a protogalactic nebula pinch plasma into the building materials of the sun and other stars. Once the material is initially pinched, gravitation will draw some of it together, especially slower-moving dust and ice particles, which will then create a seed for the growth of a central body. Moreover, the filament's vortex motion will provide angular momentum to each of the smaller agglomerations within it, generating a new, smaller set of currents carrying filaments and a new cycle of compression that forms a solar system.

(In 1989, this hypothesis now widely accepted, was definitively confirmed when scientists observed that the rotation axes of all the stars in a given cloud are aligned with the cloud's magnetic field — clearly, a magnetic-field-controlled stellar formation.)

Alfvén's theories were, of course, rejected by the cosmologists, since they challenged not only the standard model, but even called into question the existence of black holes, which were then all the rage. He had already correctly explained the cosmic rays, not as the remnants of the big bang, but as the products of electromagnetic acceleration.

> Thus, in Alfvén and Klein's scenario, only a small part of the universe — that which we see — will have first collapsed and then exploded. Instead of coming from a singular point, the explosion comes from a vast region hundreds of millions of light-years across and takes hundreds of millions of years to develop — no "origin of the universe" is required. [62]

Whether this particular theory is shown to be correct only time will tell. The important thing, as Alfvén himself points out, is that other alternative hypotheses to the big bang are possible. Whatever happens, we are sure that the model of the universe which is finally corroborated by science will have nothing in common with a closed universe with a big bang at one end and a big crunch at the other. The discovery of the telescope in 1609 was a decisive turning-point in the history of astronomy. Since then, the horizon of the universe has been pushed further and further back. Today powerful radio telescopes probe deep into space. All the time new objects are being discovered, bigger and further away, with absolutely no end in sight. Yet man's obsession with the finite creates the persistent urge to place a "final limit" on everything. We see this same phenomenon repeated time and again in the history of astronomy.

It is ironic that, at a time when technology enables us to penetrate further than ever into the vastness of the universe, we witness a psychological regression to the mediaeval world of a finite universe, beginning with Creation and ending in the total annihilation of space, time and matter. An impassable line is drawn at this point, beyond which the human mind is not meant to enquire, since "we cannot know" what is there. It is the 20th century equivalent of the old maps, which showed the edge of the world, marked with the stern warning, "Here be Monsters."

Einstein and the Big Bang

In recent decades the prejudice has become deeply rooted that "pure" science, especially theoretical physics is the product of abstract thought and mathematical deduction alone. As Eric Lerner points out, Einstein was partly responsible for this tendency. Unlike earlier theories, such as Maxwell's laws of electromagnetism, or Newton's laws of gravity, which were firmly based on experiment, and soon confirmed by hundreds of thousands of independent observations, Einstein's theories were initially confirmed on the basis of only two — the deflection of starlight by the sun's gravitational field and a slight deviation in the orbit of Mercury.

The fact that relativity theory was subsequently shown to be correct has led others, possibly not quite up to Einstein's level of genius, to assume that this is the way to proceed. Why bother with time-consuming experiments and tedious observations? Indeed, why depend upon the evidence of the senses at all, when we can get straight to the truth through the method of pure deduction?

We see a steadily increasing tendency towards a purely abstract theoretical approach to cosmology, based almost exclusively on mathematical calculations and relativity theory. "The annual number of cosmology papers published skyrocketed from sixty in 1965 to over five hundred in 1980, yet this growth was almost solely in purely theoretical work: by 1980 roughly 95 percent of these papers were devoted to various mathematical models, such as the 'Bianchi type XI universe.' By the mid-seventies, cosmologists' confidence was such that they felt able to describe in intimate detail events of the first one-hundredth second of time, several billion years ago. Theory increasingly took on the characteristic of myth — absolute, exact knowledge about events in the distant past but an increasingly hazy understanding of how they led to the cosmos we now see, and an increasing rejection of observation."

The Achilles' heel of Einstein's static, closed universe is that it would inevitably collapse in on itself because of the force of gravity. In order to get round this problem, he advanced the hypothesis of the "cosmological constant," a repulsive force which would counteract that of gravity, thus preventing the universe from collapse. For a time the idea of a static universe, held forever in a state of equilibrium by the twin forces of gravity and the "cosmological constant" received support — at least from the very small number of scientists who claimed to understand the extremely abstract and complicated theories of Einstein.

In 1970, in an article in *Science*, Gerard de Vaucouleur showed that, as objects in the universe get larger, so their density gets less. An object ten times bigger, for example, will be 100 time less dense. This has serious implications for the attempts to establish the average density of the universe, which is necessary to obtain in order to establish whether there is enough gravity to halt the Hubble Expansion. If the average density drops with increases in size, it will be impossible to define the average density for the universe as a whole. If De Vaucouleur is right, the density of the observed universe will be far less than had been thought, and the value of omega could be as little as .0002. In a universe with so little matter, the effects of gravity will be so weak that the difference between general relativity and Newtonian gravity will be insignificant and, therefore, "for all practical purposes, general relativity, the foundation of conventional cosmology, can be ignored!" Lerner continues: "De Vaucouleur's discovery shows that nowhere in the universe — except perhaps near a few ultra-dense neutron stars — is general relativity more than a subtle correction." [63]

The difficulties involved in grasping what Einstein "really meant" are proverbial. There is a story that, when some journalist asked the English scientist Eddington if it was true that there were only three people in the world who understood relativity, the latter replied, "Oh, really? And who is the third?" However, the Russian mathematician Alexander Friedmann in the early 1920s showed that Einstein's model of the universe was only one of an infinite number of possible cosmologies, some expanding, some contracting, depending on the value of the cosmological constant, and the "initial conditions" of the universe. This was a purely mathematical result, derived from Einstein's equations. The real significance of Friedmann's work was that it called into question the idea of a closed static universe, and showed that other models were possible.

Neutron Stars

Contrary to the idea of Antiquity that the stars were eternal and changeless, modern astronomy has shown that stars and other heavenly bodies have a history, a birth and life and a death — gigantic, rarefied and red in youth; blue, hot and radiant in middle life; shrunken, dense and red once more in old age. An immense amount of information has been accumulated from astronomical observations involving powerful telescopes. At Harvard alone, a quarter of a million stars had already been arranged in forty classes before the Second World War through the work of Annie J. Cannon. Now a great deal more is known as a result of radio telescopes and space exploration.

The British astronomer Fred Hoyle has made a detailed investigation of the life and death of stars. The stars are fuelled by the fusion of hydrogen into helium at the core. A star in its early stages changes little in size or temperature. This is the present position of our own sun. Sooner or later, however, the hydrogen which is being consumed at the hot center is turned into helium. This accumulates at the core until, when it reaches a certain size, quantity changes into quality. A dramatic change occurs, causing a sudden variation in size and temperature. The star expands enormously, while its surface loses heat. It become a red giant.

According to this theory, the helium core contracts, raising the temperature to the point where the helium nuclei can fuse to form carbon, releasing new energy. As it heats, it contracts still further. At this stage, the life of the star draws rapidly to a close, for the energy produced by helium fusion is far less than that produced by hydrogen fusion. At a given point, the energy required to keep the star's expansion against the pull of its own gravitational field begins to fail. The star contracts rapidly, collapsing in on itself to become a white dwarf, surrounded by a halo of gas, the remnant of the outer layers blown out by the heat of contraction. These are the basis of planetary nebulae. Stars may remain in this state for a long time, slowly cooling, until they no longer possess enough heat to glow. They then end up as black dwarves.

However, such processes seem relatively sedate in comparison to the scenario mapped out by Hoyle for bigger stars. When a large star reaches a late stage of development, in which its internal temperature reaches 3-4 billion degrees, iron begins to form at the core. At a certain stage, the temperature reaches such a point that the iron atoms are driven apart to form helium. At this point, the star collapses in on itself in about one second. Such a terrific collapse causes a violent explosion which blasts all the outer material away from the star's center. This is what is known as a supernova, like the one that astonished Chinese astronomers in the 11th century.

The question arises of what happens if a large star continues to collapse inwards under the pressure of its own gravity. Unimaginable gravitational forces would squeeze the electrons into the space already occupied by protons. According to a law of quantum mechanics known as the Pauli exclusion principle, no two electrons can occupy the same energy state in an atom. It is this principle, acting on the neutrons, that prevents further collapse. At this stage, the star is now mainly composed of neutrons, hence its name. Such a star has a tiny radius, maybe only 10 km, or about 1/700th of the radius of a white dwarf, and with a density of more than a 100 million times that of the latter, which was already extremely high. A single matchbox full of such material would weigh as much as an asteroid a mile in diameter.

With such staggering concentrations of mass, the gravitational pull of a neutron star would absorb everything in the surrounding space. The existence of such neutron stars was theoretically predicted in 1932 by the Soviet physicist Lev Landau, and later studied in detail by J. R. Oppenheimer and others. For some time it was doubted whether such stars could exist. However, in 1967 the discovery of pulsars inside the remnants of supernova such as the Crab Nebula gave rise to the theory that pulsars are really neutron stars. There is nothing in all this which is inconsistent with the principles of materialism.

Pulsars are pulsating stars which gave out rapid bursts of energy at regular intervals. It is estimated that there may be 100,000 pulsars in our galaxy alone, of which hundreds have already been located. The source of these powerful radio waves was thought to be a neutron star. According to the theory, it would have to have an immensely strong magnetic field. In the grip of the neutron star's gravitational field, electrons could only emerge at the magnetic poles, losing energy in the form of radio waves in the process. The short bursts of radio waves could be explained by fact that the neutron star must be rotating. In 1969, it was noted that a light of a dim star in the Crab Nebula was flashing intermittently in line with the microwave pulses. This was the first sighting of a neutron star. Then, in 1982, a fast pulsar was discovered, with pulsations 20 times faster than those of the Crab Nebula — 642 times a second. In the 1960s, new objects were discovered by radio telescopes, the quasars. By the end of the decade, 150 were discovered — some of them estimated to be nine billion light years away, assuming the red-shift to be correct. To appear at all at such a vast distance must mean that these objects are 30 to 100 times more luminous than a normal galaxy. Yet they appeared to be small. This poses difficulties, which led some astronomers to refuse to accept that they could be so far away.

The discovery of quasars gave an unexpected boost to the big bang theory. The existence of collapsed stars with a tremendously strong gravitational field posed problems which could not be resolved by direct observation. This fact opened the door to a flood of speculations, involving the most peculiar interpretations of Einstein's general theory of relativity. As Eric Lerner points out:

> The glamour of the mysterious quasars quickly attracted young researchers to the arcane calculations of general relativity and thus to cosmological problems, especially those of a mathematical nature. After 1964 the number of papers published in cosmology leapt upward, but the growth was almost wholly in purely theoretical pieces — mathematical examinations of some problem in general relativity, which made no effort to compare results with observations. Already, in 1964, perhaps four out of five cosmology papers were theoretical, where only a third had been so a decade earlier. [64]

It is necessary to distinguish clearly between black holes, the existence of which has been derived from a particular interpretation of the general theory of relativity, and neutron stars, which have actually been observed. The idea of black holes has captured the imagination of millions through the writings of authors like Stephen Hawking. Yet the existence of block holes is not universally accepted, nor has it been definitively proved. Roger Penrose, in an essay based on a BBC Radio lecture delivered in 1973, describes the theory of black holes as follows:

> What is a black hole? For astronomical purposes it behaves as a small, highly condensed dark "body." But it is not really a material body in the ordinary sense. It possesses no ponderable surface. A black hole is a region of empty space (albeit a strangely distorted one) which acts as a center of gravitational attraction. At one time a material body was there. But the body collapsed inwards under its own grav-itational pull. The more the body concentrated itself towards the center the stronger became its gravitational field and the less was the body able to stop itself from yet further collapse. At a certain stage a point of no return was reached, and the body passed within its "absolute event horizon."
>
> I shall say more of this later, but for our present purposes, it is the absolute event horizon which acts as the boundary surface of the black hole. This surface is not material. It is merely a demarcation line drawn in space separating an interior from an exterior region. The interior region — into which the body has fallen — is defined by the fact that no matter, light, or signal of any kind can escape from it, while the exterior region is where it is still possible for signals or material particles to escape to the outside world. The matter which collapsed to form the black hole has fallen deep inside to attain incredible densities, apparently even to be crushed out of existence by reaching what is known as a 'space-time singularity' — a place where physical laws, as presently understood, must cease to apply. [65]

Stephen Hawking

In 1970, Stephen Hawking put forward the idea that the energy content of a black hole might occasionally produce a pair of subatomic particles, one of which might escape. This implies that a black hole can evaporate, although this would take an unimaginably long period of time. In the end, according to this view, it would explode, producing a large amount of gamma rays. Hawking's theories have attracted a lot of attention. His well-written best-seller *A Brief History of Time, From the Big Bang to Black Holes*, was perhaps the book that more than any other drew the attention of the new theories of cosmology to the public's attention. The author's lucid style made complicated ideas seem both simple and attractive. It makes for good reading, but so do many works of science fiction. Regrettably, it appears to have become fashionable for the authors of popular works about cosmology to sound as mystical as possible, and to put forward the most outlandish theories, based on the maximum amount of speculation and the minimum amount of facts. Mathematical models have displaced observation almost entirely. The central philosophy of this school of thought is summed up in Stephen Hawking's aphorism "one cannot really argue with a mathematical theorem."

Hawking claims that he and Roger Penrose proved (mathematically) that the general theory of relativity "implied that the universe must have a beginning and, possibly, an end." The basis of all this is that the general theory of relativity is taken as absolutely true. Yet, paradoxically, at the point of the big bang general relativity suddenly becomes irrelevant. It ceases to apply, just as all the laws of physics cease to apply, so that nothing whatsoever can be said about it. Nothing, that is, except metaphysical speculation of the worst sort. But we will return to this later.

According to this theory, time and space did not exist before the big bang, when all the matter in the universe was alleged to have been concentrated at a single infinitesimally small point, known to mathematicians as a singularity. Hawking himself points out the dimensions involved in this remarkable cosmological transaction:

> We now know that our galaxy is only one of some hundred thousand million that can be seen using modern telescopes, each galaxy itself containing some hundred thousand million stars...We live in a galaxy that is about one hundred thousand light-years across and is slowly rotating; the stars in its spiral arms orbit around its center about once every several hundred million years. Our sun is just an ordinary, average-sized, yellow star, near the inner edge of one of the spiral arms. We have certainly come a long way since Aristotle and Ptolemy, when we thought that the earth was the center of the universe! [66]

In point of fact, the very large quantities of matter mentioned here give no real idea of the amount of matter in the universe. New galaxies and super-clusters are being discovered all the time, and there is no end to this process. We may have come a long way since Aristotle in some respects. But in others, it seems that we are far, far behind him. Aristotle would never have made the mistake of talking about a time before time existed, or claiming that the entire universe was, in effect, created from nothing. In order to find ideas like these one would have to go back several thousand years to the world of the Judaic-Babylonian Creation myth.

Whenever someone attempts to protest against these proceedings, he is instantly ushered into the presence of the great Albert Einstein, as a naughty schoolboy is dragged to the headmaster's study, and given a stern lecture on the need to show greater respect to general relativity, informed that one cannot argue with mathematical theorems, and sent home duly chastened. The main difference is that most headmasters are alive, and Einstein is dead, and therefore unable to comment on this particular interpretation of his theories. In fact, one would look in vain in all the writings of Einstein for any reference to the big bang, black holes and the like. Einstein himself, although he initially tended towards philosophical idealism, was implacably opposed to mysticism in science. He spent the last decades of his life fighting against the subjective idealist views of Heisenberg and Bohr, and, in fact, moved close to a materialist position. He would certainly have been horrified that mystical conclusions should be drawn from his theories. The following is a good example:

> All of the Friedmann solutions have the feature that at some time in the past (between ten and twenty thousand million years ago) the distance between neighboring galaxies must have been zero. At that time, which we call the big bang, the density of the universe and the curvature of space-time would have been infinite. Because mathematics cannot really handle infinite numbers, this means that the general theory of relativity (on which Friedmann's solutions are based) predicts that there is a point in the universe where the theory itself breaks down. Such a point is an example of what mathematicians call a singularity. In fact, all our theories of science are formulated on the assumption that space-time is smooth and nearly flat, so they break down at the big bang singularity, where the curvature of space-time is infinite. This means that even if there were events before the big bang, one could not use them to determine what would happen afterward, because predictability would break down at the big bang. Correspondingly, if, as is the case, we know only what has happened since the big bang, we could not determine what happened beforehand. As far as we are concerned, events before the big bang can have no consequences, so they should therefore cut them out of the model and say that time had a beginning at the big bang.

Passages such as this forcefully remind one of the intellectual gymnastics of the medieval Schoolmen, arguing about the number of angels who could dance on the end of a pin. This is not meant as an insult. If the validity of an argument is determined by its internal consistency, then the arguments of the Schoolmen were as valid as this. They were not fools, but highly skilled logicians and mathematicians, who erected theoretical constructs as elaborate and perfect in their way as medieval cathedrals. All that was necessary was to accept their premises, and everything fell into place. The problem is whether the original premise is valid or not. This is a general problem with all mathematics, and its central weakness. And this entire theory leans very heavily on mathematics.

"At the time which we call the big bang..." — But if there was no time, how can we refer to it as "a time" at all? Time is said to have begun at that point. So what was there before time? A time when there was no time! The self-contradictory nature of this idea is glaringly obvious. Time and space are the mode of existence of matter. If there was neither time, nor space, nor matter, what was there? Energy? But energy, as Einstein explains, is just another manifestation of matter. A force-field? But a force-field is also energy, so the difficulty remains. The only way that time can be got rid of, is if before the big bang there was — nothing.

The problem is: how is it possible to get from nothing to something? If one is religiously minded, there is no problem; God created the universe from nothing. This is the doctrine of the Catholic Church, of Creation ex nihilo. Hawking is uncomfortably aware of this fact, as he says in the very next line:

"Many people do not like the idea that time has a beginning, probably because it smacks of divine intervention. (The Catholic Church, on the other hand, seized on the big bang model and in 1951 officially pronounced it to be in accordance with the Bible.)"

Hawking himself does not want to accept this conclusion. But it is unavoidable. The whole mess arises out of a philosophically incorrect concept of time. Einstein was partly responsible for this, since he appeared to introduce a subjective element by confusing the measurement of time with time itself. Here again the reaction against the old mechanical physics of Newton has been carried to an extreme. The question is not whether time is "relative" or "absolute." The central issue to be addressed is whether time is objective or subjective; whether time is the mode of existence of matter or an entirely subjective concept existing in the mind and determined by the observer. Hawking clearly adopts a subjective view of time, when he writes:

"Newton's laws of motion put an end to the idea of absolute position in space. The theory of relativity gets rid of absolute time. Consider a pair of twins. Suppose that one twin goes to live on the top of a mountain while the other stays at sea level. The first twin would age faster than the second. Thus, if they met again, one would be older than the other. In this case, the difference in ages would be very small, but it would be much larger if one of the twins went for a long trip in a spaceship at nearly the speed of light. When he returned, he would be much younger than the one who stayed on Earth. This is known as the twins paradox, but it is a paradox only if one has the idea of absolute time at the back of one's mind. In the theory of relativity there is no unique absolute measure of time that depends on where he is and how he is moving." [67]

That there is a subjective element in the measurement of time is not in dispute. We measure time according to a definite frame of reference, which can, and does, vary from one place to another. The time in London is different from the time in Sydney or New York. But this does not mean that time is purely subjective. The objective processes in the universe take place whether we are able to measure them or not. Time, space, and motion are objective to matter, and have no beginning and no end.

Here it is interesting to note what Engels had to say on the subject:

Let us continue. So time had a beginning. What was there before this beginning? The universe, which was then in a self-identical, unchanging state. And as no changes succeed one another in this state, the more specialized idea of time transforms itself into the more general idea of being. In the first place, we are not in the least concerned here with what concepts change in Herr Dühring's head. The subject at issue is not the concept of time, but real time, which Herr Dühring will by no means rid himself of so cheaply. In the second place, however much the concept of time may be converted into the more general idea of being, this takes us not one step further. For the basic forms of all being are space and time, and being out of time is just as gross an absurdity as being out of space.

The Hegelian "timelessly past being" and the neo-Schellingian "unpreconceivable being" are rational ideas compared with this being out of time. For this reason Herr Dühring sets to work very cautiously; actually it is of course time, but of such a kind as cannot really be called time; time does not in itself consist of real parts, and is only divided up arbitrarily by our understanding — only an actual filling of time with differentiable facts is susceptible of being counted — what the accumulation of empty duration means is quite unimaginable. What this accumulation is supposed to mean is immaterial here; the question is whether the world, in the state assumed here, has duration, passes through a duration in time. We have long known that we can get nothing by measuring such a duration without content, just as we can get nothing by measuring without aim or purpose in empty space; and Hegel calls this infinity bad precisely because of the tedium of this procedure. [68]

Do Singularities Exist?

A black hole and a singularity are not the same thing. There is nothing in principle which excludes the possible existence of stellar black holes, in the sense of a massive collapsed star where the force of gravity is so immense that not even light can escape from its surface. Even the idea is not new. It was predicted in the 18th century by John Mitchell who pointed out that a sufficiently massive star would trap light. He came to this conclusion on the basis of Newton's classical theory of gravitation. General relativity did not enter into it.

However, the theory advanced by Hawking and Penrose goes far beyond the observed facts, and, as we have seen, draws conclusions which lend themselves to all kinds of mysticism, even if this was not their intention. Eric Lerner considers the case for supermassive black holes at the center of galaxies to be weak. Together with Anthony Peratt, he has shown how all the features associated with these supermassive black holes, quasars, etc., can be better explained by electromagnetic phenomena. However, he believes the evidence is considerably stronger for the existence of stellar sized black holes since this rests on detecting very intense X-ray sources which are too big to be neutron stars. But even here the observations are far from proving the case.

The abstractions of mathematics are useful tools for understanding the universe, on one condition: that we do not lose sight of the fact that even the best mathematical model is only a rough approximation of reality. The problems start when people begin confusing the model with the thing itself. Hawking himself unwittingly reveals the weakness of this method in the passage already quoted. He assumes that the density of the universe at the point of the big bang was infinite, without giving any reasons for this, and then adds, in a most peculiar line of argument, that "because mathematics cannot really handle infinite numbers" the theory of relativity breaks down at this point. To this, it is necessary to add, "and all the known laws of physics," since it is not only general relativity which breaks down with the big bang, but all of science. It is not just that we do not know what occurred before this. It is that we cannot know.

This is a return to Kant's theory of the unknowable Thing-in-Itself. In the past, it was the role of religion and certain idealist philosophers, like Hume and Kant, to place a limit upon human understanding. Science was permitted to go so far, and no further. At the point where human intelligence was not allowed to proceed, mysticism, religion and irrationality commenced. Yet the whole history of science is the story of how one barrier after another was removed. What was supposed to be unknowable for one generation became an open book for the next. The whole of science is based on the notion that the universe can be known. Now, for the first time, scientists are placing limits on knowledge, an extraordinary state of affairs and a sad comment on the present situation in theoretical physics and cosmology.

Consider the implications of the above passage: a) since the laws of science, including general relativity (which is supposed to provide the basis for the whole theory) break down at the big bang, it is impossible to know what, if anything, occurred before it, b) even if there were events before the big bang, they have no relevance to what happened afterwards, c) we cannot know anything about it, and so, d) we should simply "cut it out of the model and say that time began at the big bang."

The self-assurance with which these assertions are put forward is truly breathtaking. We are asked to accept an absolute limit on our ability to understand the most fundamental problems in cosmology, in effect, to ask no questions (because all questions about the time before there was time are meaningless) and that we should just accept without more ado that time began with the big bang. In this way, Stephen Hawking simply assumes what has to be proved. In the same way, the theologians assert that God created the universe, and when asked who created God, merely answer that such questions are beyond the minds of mortals. On one thing we can agree, however; the whole thing does indeed "smack of divine intervention." More than that, it necessarily implies it.

In his polemic against Dühring, Engels points out that it is impossible that motion should arise out of immobility, that something should arise out of nothing: "Without an act of Creation we can never get from nothing to something, even if the something were as small as a mathematical differential." [69] Hawking's principal defense seems to be that the alternative theory to the big bang, put forward by Fred Hoyle, Thomas Gould and Hermann Bondi — the so-called Steady State theory — was shown to be false. From the standpoint of dialectical materialism, there was never anything to choose between these two theories. One was as bad as the other. Indeed the Steady State theory, which suggested that matter was being continuously created in space out of nothing, was, if possible, even more mystical than its rival. The very fact that such an idea could be taken seriously by scientists is itself a damning comment on the philosophical confusion which has bedeviled science for so long.

The ancients already understood that "out of nothing comes nothing." This fact is expressed in one of the most fundamental laws of physics, the law of the conservation of energy. Hoyle's claim that only a very small amount was involved makes no difference. It is a bit like the naïve young lady who, in order to placate her irate father who found out she was going to have a baby, assured him that it was "only a little one." Not even the tiniest particle of matter (or energy, which is the same) can ever be created or destroyed, and therefore the Steady State theory was doomed from the outset.

Penrose's theory of a "singularity" was originally nothing to do with the origin of the universe. It merely predicted that a star collapsing under its own gravity would be trapped in a region whose surface eventually shrinks to zero size. In 1970, however, he and Hawking produced a joint paper in which they claimed to prove that the big bang itself was such a "singularity," provided only that "general relativity is correct and the universe contains as much matter as we observe."

> There was a lot of opposition to our work, partly from the Russians because of their Marxist belief in scientific determinism, and partly from people who felt that the whole idea of singularities was repugnant and spoiled the beauty of Einstein's theory. However, one cannot really argue with a mathematical theorem. So in the end our work became generally accepted and nowadays nearly everyone assumes that the universe started with a big bang singularity.

General relativity has proved a very powerful tool, but every theory has its limits, and one has the impression that it is being pushed to the limit here. How long it will be before it is replaced by a broader and more comprehensive set of ideas it is impossible to say, but it is clear that this particular application of it has led to a blind alley. As far as the amount of matter in the universe is concerned, the total amount will never be known, because it has no limit. Typically, they are so wrapped up in mathematical equations, that they forget reality. In practice, the equations have replaced reality.

Having succeeded in convincing a lot of people, on the basis that "one cannot really argue with a mathematical theorem," Hawking then proceeded to have second thoughts: "It is perhaps ironic," he says, "that, having changed my mind, I am now trying to convince other physicists that there was in fact no singularity at the beginning of the universe — as we shall see later, it can disappear once quantum effects are taken into account." The arbitrary nature of the whole method is shown in Hawking's extraordinary change of mind. He now says there is no singularity in the big bang. Why? What has changed? There is no more actual evidence than before. These twists and turns all take place in the world of mathematical abstractions.

Hawking's theory of black holes represents an extension of the idea of singularity to particular parts of the universe. It is full of the most contradictory and mystical elements. Take the following passage, which describes the extraordinary scenario of an astronaut falling into a black hole:

"The work that Roger Penrose and I did between 1965 and 1970 showed that, according to general relativity, there must be a singularity of infinite density and space-time curvature within a black hole. This is rather like the big bang at the beginning of time, only it would be an end of time for the collapsing body and the astronaut. At this singularity the laws of science and our ability to predict the future would break down. However, any observer who remained outside the black hole would not be affected by this failure of predictability, because neither light nor any other signal could reach him from the singularity. This remarkable fact led Roger Penrose to propose the cosmic censorship hypothesis, which might be paraphrased as 'God abhors a naked singularity.' In other words, the singularities produced by gravitational collapse occur only in places, like black holes, where they are decently hidden from outside view by an event horizon. Strictly, this is what is known as the weak cosmic censorship hypothesis: it protects observers who remain outside the black hole from the consequences of the breakdown of predictability that occurs at the singularity, but it does nothing at all for the poor unfortunate astronaut who falls into the hole." [70]

What sense can one make of this? Not content with the beginning (and end) of time for the universe as a whole, Penrose and Hawking now discover numerous parts of the universe where time has already ended! Although the evidence for the existence of black holes is patchy, it seems likely that some such phenomenon does exist, in the form of collapsed stars with tremendous concentrations of matter and gravity. But it seems extremely doubtful that this gravitational collapse could ever reach the point of a singularity, much less remain in this state forever. Long before this point was reached, such a tremendous concentration of matter and energy would result in a massive explosion.

The entire universe is proof that the process of change is never-ending, at all levels. Vast tracts of the universe may be expanding, while others are contracting. Long periods of apparent equilibrium are disrupted by violent explosions, like supernovas, which in turn provide the raw material for the formation of new galaxies, which goes on all the time. There is no disappearance or creation of matter, but only its continuous, restless change from one state to another. There can therefore be no question of the "end of time" inside a black hole, or anywhere else.

An Empty Abstraction

The whole mystical notion derives from the subjectivist interpretation of time, which makes it dependent on ("relative to") an observer. But time is an objective phenomenon, which is independent of any observer. The need to introduce the unfortunate astronaut into the picture does not arise from any scientific necessity, but is the product of a definite philosophical point of view, smuggled in under the banner of "relativity theory." You see, for time to be "real," it needs an observer, who can then interpret it from his or her point of view. Presumably, if there is no observer, there is no time! In a most peculiar piece of reasoning, this observer is protected against the malign influence of the black hole, by an arbitrary hypothesis, a "weak cosmic censorship," whatever that might mean. Inside the hole, however, there is no time at all. So outside, time exists, but a little distance away, time does not exit. At the boundary between the two states, we have the mysterious event horizon, the nature of which is shrouded in obscurity.

At least, it would appear that we must abandon all hope of ever understanding what goes on beyond the event horizon, since, to quote Hawking, it is "decently hidden from outside view." Here we have the 20th century equivalent of the Kantian Thing-in-Itself. And, like the Thing-in-Itself, it turns out to be not so difficult to understand after all. What we have here is a mystical idealist view of time and space, fed into a mathematical model, and mistaken for something real.

Time and space are the most fundamental properties of matter. More correctly, they are the mode of existence of matter. Kant already pointed out that, if we leave aside all the physical properties of matter, we are left with time and space. But this is, in fact, an empty abstraction. Time and space can no more exist separately from the physical properties of matter than one can consume "fruit" in general, as opposed to apples and oranges, or make love to Womankind. The accusation has been leveled against Marx, without the slightest justification, that he conceived of History as taking place without the conscious participation of men and women, as a result of Economic Forces, or some nonsense of the sort. In fact, Marx states quite clearly that History can do nothing, and that men and women make their own history, although they do not do so entirely according to their own "free will."

Hawking, Penrose and many others are guilty precisely of the mistake that was falsely attributed to Marx. Instead of the empty abstraction History, which is, in effect, personified, and endowed with a life and a will of its own, we have the equally empty abstraction Time, envisaged as an independent entity which is born and dies, and generally gets up to all kinds of tricks, along with its friend, Space, which arises and collapses and bends, a bit like a cosmic drunkard, and ends up swallowing hapless astronauts in black holes.

Now this kind of thing is fine in science fiction, but is not very useful as a means of understanding the universe. Clearly, there are immense practical difficulties in obtaining precise information about, say, neutron stars. In a sense, in relation to the universe, we find ourselves in a position roughly analogous to early humans in relation to natural phenomena. Lacking adequate information, we seek a rational explanation of difficult and obscure things. We are thrown back on our own resources — the mind and the imagination. Things seem mysterious when they are not understood. In order to understand, it is necessary to make hypotheses. Some of these will be found to be wrong. That, in itself, presents no problem. The whole history of science is full of examples where the pursuit of an incorrect hypothesis led to important discoveries.

However, we have a duty to attempt to ensure that hypotheses have a reasonably rational character. Here the study of philosophy becomes indispensable. Do we really have to go back to primitive myths and religion in order to make sense of the universe? Do we need to revive the discredited notions of idealism, which, in fact, are closely related to the former? Is it really necessary to re-invent the wheel? "One cannot argue with a mathematical theorem." Maybe not. But it is certainly possible to argue with false philosophical premises, and an idealist interpretation of time, which leads us to conclusions like the following:

> There are some solutions of the equations of general relativity in which it is possible for our astronaut to see a naked singularity: he may be able to avoid hitting the singularity and instead fall through a "wormhole" and come out in another region of the universe. This would offer great possibilities for travel in space and time, but unfortunately it seems that these solutions may all be highly unstable; the least disturbance, such as the presence of an astronaut, may change them so that the astronaut could not see the singularity until he hit it and his time came to an end. In

other words, the singularity would always lie in his future and never in his past. The strong version of the cosmic censorship hypothesis states that in a realistic solution, the singularities would always lie either entirely in the future (like the singularities of gravitational collapse) or entirely in the past (like the big bang). It is greatly to be hoped that some version of the censorship hypothesis holds because close to naked singularities it may be possible to travel into the past. While this would be fine for writers of science fiction, it would mean that no one's life would ever be safe: someone might go into the past and kill your father or mother before you were conceived! [71]

"Time-travel" belongs to the pages of science fiction, where it can be a source of harmless amusement. But we are convinced that nobody ought to be afraid that their existence may be put at risk by some inconsiderate time-traveler doing away with their granny. Frankly, one only has to pose the question to realize that it is a patent absurdity. Time moves in only one direction, from past to future, and cannot be reversed. Whatever our friend the astronaut might find at the bottom of a black hole, he will not find that time has been reversed, or "stands still" (except in the sense that, since he would instantly be torn to pieces by the force of gravity, time would cease for him, along with a lot of other things).

We have already commented on the tendency to confuse science with science fiction. It is also noticeable that much of science fiction itself is permeated with a semi-religious, mystical and idealist spirit. Long ago, Engels pointed out that scientists who despised philosophy frequently fall victim to all kinds of mysticism. He wrote an article on the subject entitled *Natural Science and the Spirit World*, from which the following extract is taken:

> This school prevails in England. Its father, the much lauded Francis Bacon, already advanced the demand that his new empirical, inductive method should be pursued to attain, above all, by its means: longer life, rejuvenation — to a certain extent, alteration of stature and features, transformation of one body into another, the production of new species, power over the air and the production of storms. He complains that such investigations have been abandoned, and in his natural history he gives definite recipes for making gold and performing various miracles. Similarly Isaac Newton in his old age greatly busied himself with expounding the Revelation of St. John. So it is not to be wondered at if in recent years English empiricism in the person of some of its representatives — and not the worst of them — should seem to have fallen a hopeless victim to the spirit-rapping and spirit-seeing imported from America. [72]

There is no doubt that Stephen Hawking and Roger Penrose are brilliant scientists and mathematicians. The problem is that, if you begin with a wrong premise, you will inevitably draw the wrong conclusions. Hawking clearly feels uncomfortable with the idea that religious conclusions can be drawn from his theories. He mentions that in 1981 he attended a conference on cosmology in the Vatican, organized by the Jesuits, and comments:

> The Catholic Church had made a bad mistake with Galileo when it tried to lay down the law on a question of science, declaring that the sun went round the earth. Now, centuries later, it had decided to invite a number of experts to advise it on cosmology. At the end of the conference the participants were granted an audience with the Pope. He told us that it was all right to study the evolution of the universe after the big bang, but we should not inquire into the big bang itself because that was the moment of Creation and therefore the work of God. I was glad then that he did not know the subject of the talk I had just given at the conference — the possibility that space-time was finite but had no boundary, which means that it had no beginning, no moment of Creation. I had no desire to share the fate of Galileo, with whom I feel a strong sense of identity, partly because of the coincidence of having been born exactly 300 years after his death! [73]

Clearly, Hawking wishes to draw a line between himself and the Creationists. But the attempt is not very successful. How can the universe be finite, and yet have no boundaries? In mathematics, it is possible to have an infinite series of numbers which starts with one. But in practice, the idea of infinity cannot begin with one, or any other number. Infinity is not a mathematical concept. It cannot be counted. This one-sided "infinity" is what Hegel calls bad infinity. Engels deals with this question in his polemic with Dühring:

> But what of the contradiction of "the counted infinite numerical series"? We shall be in a position to examine it more closely a soon as Herr Dühring has performed the clever trick of counting it for us. When he has completed the task of counting from − _ (minus infinity) to 0, let him come again. It is certainly obvious that, wherever he begins to count, he will leave behind him an infinite series and, with it, the task which he has to fulfill. Just let him invert his own infinite series 1+2+3+4...and try to count from the infinite end back to 1; it would obviously only be attempted by a man who has not the faintest understanding of what the problem is. Still more. When Herr Dühring asserts that the infinite series of lapsed time has been counted, he is thereby asserting that time has a beginning; for otherwise he would have been unable to start "counting" at all. Once again, therefore, he smuggles into the argument, as a premise, what he has to prove. The idea of an infinite series which has been counted, in other words, the world-encompassing Dühringian Law of Determinate Number, is therefore a contradiction in adjecto, contains within itself a contradiction, and indeed an absurd contradiction.

It is clear that an infinity which has an end but no beginning is neither more nor less infinite than one with a beginning but no end. The slightest dialectical insight should have told Herr Dühring that beginning and end necessarily belong together, like the North Pole and the South Pole, and that if the end is left out, the beginning just becomes the end — the one end which the series has; and vice versa. The whole deception would be impossible but for the mathematical usage of working with infinite series. Because in mathematics it is necessary to start from determinate, finite terms in order to reach the indeterminate, the infinite, all mathematical series, positive or negative, must start with 1, or they cannot be used for calculation. But the logical need of the mathematician is far from being a compulsory law for the real world. [74]

Stephen Hawking carried this relativistic speculation to an extreme with his work on black holes, which leads us right into the realms of science fiction. In an attempt to get round the awkward question of what happened before the big bang, the idea was advanced of "baby universes," coming into existence all the time, and connected by so-called "wormholes." As Lerner ironically comments: "It is a vision that seems to beg for some form of cosmic birth control." [75] It really is astounding that sober scientists could take such grotesque ideas for good coin.

The idea of a "finite universe with no boundaries" is yet another mathematical abstraction, which does not correspond to the reality of an eternal and infinite, constantly changing universe. Once we adopt this standpoint, there is no need for mystical speculations about "wormholes," singularities, superstrings, and all the rest of it. An infinite universe does not require us to look for a beginning or an end, only to trace the endless process of movement, change and development. This dialectical conception leaves no room for Heaven or Hell, God or the Devil, Creation or the Last Judgment. The same cannot be said for Hawking who, quite predictably, ends up attempting to "know the mind of God."

The reactionaries rub their hands at this spectacle, and use the prevailing current of obscurantism in science for their own ends. William Rees-Mogg, big business consultant, writes:

> We think it is extremely likely that the religious movement we see at work in many societies across the globe will be strengthened if we go through a very difficult economic period. Religion will be strengthened because the current thrust of science no longer undermines the religious perception of reality. Indeed, for the first time in centuries, it actually buttresses it. [76]

Thoughts in a Vacuum

Why, sometimes, I've believed as many as six impossible things before breakfast. (Lewis Carroll)

"With men this is impossible;
but with God all things are possible. (Matthew, 19:26)

Nothing can be created out of nothing. (Lucretius)

Just before finishing writing this book, we came across the latest contribution to cosmology of the big bang, which appeared in *The New Scientist* on February 25, 1995. In an article by Robert Matthews entitled *Nothing like a Vacuum*, we read the following: "It is all around you, yet you cannot feel it. It is the source of everything, yet is nothing."

What is this amazing thing? A vacuum. What is a vacuum? The Latin word *vacuus*, from which it comes, means quite simply empty. The dictionary defines it as "space empty, or devoid of all matter or content; any space unoccupied or unfilled; a void, blank." This was the case up till now. But not any longer. The humble vacuum, in Mr. Matthews words, has become "one of the hottest topics in contemporary physics."

"It is proving to be a wonderland of magical effects: force fields that emerge from nowhere, particles popping in and out of existence and energetic jitterings with no apparent power source."

Thanks to Heisenberg and Einstein (poor Einstein!), we have the "astonishing realization that all around us 'virtual' subatomic particles are perpetually popping up out of nothing, and then disappearing again within about 10–23 seconds. 'Empty space' is thus not really empty at all, but a seething sea of activity that pervades the entire Universe." This is true and false. It is true that the whole universe is pervaded by matter and energy, and that "empty space" is not really empty, but full of particles, radiation and force-fields. It is true that particles are constantly changing, and that some have a life so fleeting that they are called "virtual" particles. There is absolutely nothing "astonishing" about these ideas, which were known decades ago. But it is entirely untrue that they pop "out of nothing." We have already dealt with this misconception above, and it is not necessary to repeat what was said. Like an old record with a repeating groove, those who wish to introduce idealism into physics constantly harp on the idea that you can get something from nothing. This idea contradicts all the known laws of physics, including quantum physics. Yet we find here the incredible notion that energy can be obtained literally from nothing! This is like the attempts to discover perpetual motion, which were rightly ridiculed in the past.

Modern physics begins with the rejection of the old idea of the ether, an invisible universal medium, through which light-waves were thought to travel. Einstein's theory of special relativity proved that light could travel through a vacuum, and did not require any special medium. Incredibly, after citing Einstein as an authority (as obligatory nowadays as crossing yourself before leaving church, and about as meaningful) Mr. Matthews proceeds to smuggle the ether back into physics:

"This does not mean that a universal fluid cannot exist, but it does mean that such a fluid must conform to the dictates of special relativity. The vacuum is not forced to be mere quantum fluctuations around an average state of true nothingness. It can be a permanent, non-zero source of energy in the Universe."

Now what precisely is one supposed to make of this? So far we have been told about "astonishing" new developments in physics, "wonderlands" of particles and have been assured that vacuums possess enough energy to solve all our needs. But the actual information provided by the article does not seem to say anything new. It is very long on assertions, but very short on facts. Perhaps it was the author's intention to make up for this by obscurity of expression. What is meant by a "permanent non-zero source of energy" is anyone's guess. And what is an "average state of true nothingness"? If what is meant is a true vacuum, then it would have been preferable to use two clear words instead of four unclear ones. This kind of deliberate obscurity is generally used to cover up muddled thinking, especially in this area. Why not speak plainly? Unless, of course, what is involved is a "true nothingness" — of content.

The whole thrust of the article is to show that a vacuum derives unlimited quantities of energy from nowhere. The only "proof" for this are a couple of references to the special and general theories of relativity, which are regularly used as a peg upon which to hang any arbitrary hypothesis. "Special relativity demands that the vacuum's properties must appear the same for all observers, whatever their speed. For this to be true it turns out that the pressure of the vacuum 'sea' must exactly cancel out its energy density. It is a condition that sounds harmless enough, but it has some astounding consequences. It means, for example, that a given region of vacuum energy retains the same energy density, no matter how much the region expands. This is odd, to say the least. Compare it with the behavior of an ordinary gas, whose energy density decreases as its volume increases. It is as if the vacuum can draw on a constant reservoir of energy."

In the first place, we note that what was only a hypothetical "universal fluid" a couple of sentences ago has now become transformed into an actual vacuum "sea," though where all the "water" came from, nobody is quite sure. This is odd to say the least. But leave it there. Let us, like the author, assume what was to be proved, and accept the existence of this vast ocean of nothingness. It turns out that this "nothing" is now not only something, but a very substantial "something." As if by magic, it is filled with energy from a "constant reservoir." This is the cosmological equivalent of the cornucopia, the "horn of plenty" of Greek and Irish mythology, a mysterious drinking horn or cauldron which, however much one drank from it, was never empty. This was a gift from the gods. Now Mr. Matthews wishes to present us with something that makes this look like child's play.

If energy enters a vacuum, it must come from somewhere outside the vacuum. This is plain enough, since a vacuum cannot exist in isolation from matter and energy. The idea of empty space without matter is as nonsensical as the idea of matter without space. There is no such thing as a perfect vacuum on earth. The nearest thing to a perfect vacuum is space. But in point of fact, space is not empty, either. Decades ago, Hannes Alfvén pointed out that space was alive with networks of electrical currents and magnetic fields filled with plasma filaments. This is not the results of speculation or appeals to relativity theory, but is borne out by observation, including those of the Voyager and Pioneer spacecrafts which detected these currents and filaments around Jupiter, Saturn and Uranus.

So there is, indeed, plenty of energy in space. But not the kind of energy Mr. Matthews is talking about. Not a bit of it. Having established his "vacuum sea" he means to get his energy directly from the vacuum. No matter required! This is much better than the conjurer who pulls a rabbit out of a hat. After all, we all know the rabbit actually comes from somewhere. This energy comes from nowhere at all. It comes from a vacuum, by courtesy of the general theory of relativity: "One of the key features of Einstein's general relativity theory is that mass is not the only source of gravitation. In particular, pressure, both positive and negative can also give rise to gravitational effects."

By this point, the reader is thoroughly mystified. Now, however, all becomes clear (almost).

> This feature of the vacuum [we are now told] lies at the heart of perhaps the most important new concept in cosmology of the past decade: cosmic inflation. Developed principally by Alan Guth at MIT and Andrei Linde, now at Stanford, the idea of cosmic inflation arises from the assumption that the very early Universe was packed with unstable vacuum energy whose "antigravitational" effect expanded the Universe by a factor of perhaps 1050 in just 10–32 seconds. The vacuum energy died away, leaving random fluctuations whose energy turned into heat. Because energy and matter are interchangeable, the result was the matter creation we now call the big bang.

So that's it! The whole arbitrary construction is meant to back up the inflationary theory of the big bang. As always, they move the goalposts continually, in order to prop up their hypothesis at all costs. It is like the supporters of the old Aristotle-Ptolemaic theory of the crystal spheres, which they continually revised, making it ever more complicated, in order to fit the facts. As we have seen, the theory has been having a bad time lately, what with the missing "cold dark matter" and the unholy mess about the Hubble constant.

Badly in need of a little support, its supporters have obviously looked round for some explanation to one of the central problems of the theory — where did all the energy come from to cause the inflationary big bang. "The biggest free lunch of all time," Alan Guth called it. Now they want to pass the bill to somebody, or something, and come up with — a vacuum. We doubt whether this particular bill will ever be paid. And, in the real world, people who don't pay their bills are usually unceremoniously shown the door, even if they offer to produce the general theory of relativity in lieu of cash.

"From nothing, through nothing, to nothing," said Hegel. That is a fitting epitaph for the theory of inflation. There is actually only one way of getting something from nothing — by an act of Creation. And that is only possible through the intervention of a Creator. Try as they will, the supporters of the big bang will find that their footsteps will always lead them in this direction. Some will go quite happily, others protesting that they are not religious "in the conventional sense." But the movement back to mysticism is the inevitable consequence of this modern Creation myth. Fortunately, an increasing number of people are becoming dissatisfied with this state of affairs. Sooner or later, a breakthrough will occur at the level of observation which will enable a new theory to emerge, allowing the big bang to be laid decently to rest. The sooner the better.

The Origins of the Solar System

Space is not really empty. A perfect vacuum does not exist in nature. Space is filled by a thin gas — "interstellar gas" first detected in 1904 by Hartmann. The concentrations of gas and dust become much greater and denser in the neighborhood of galaxies, which are surrounded by "fog," mostly composed of atoms of hydrogen, ionized by radiation from the stars. Even this matter is not inert and lifeless, but is broken up into electrically-charged subatomic particles, subject to all kinds of movement, processes and change. These atoms occasionally collide and can change their energy state. Though an individual atom might only collide once every 11 million years, given the vast numbers involved, it is enough to give rise to a continuous and detectable emission, the "song of hydrogen," first detected in 1951.

Almost all of this is hydrogen, but there is also deuterium, a more complex form of hydrogen, oxygen and helium. It might seem impossible that combination should occur, given the extremely sparse distribution of these elements in space. But occur it does, and to a remarkable degree of complexity. The water molecule ($H2O$), was found in space, as was that of ammonia ($NH3$), followed by formaldehyde ($H2CO$), and even more complex molecules, giving rise to a new science — astrochemistry. Finally, it has been proved that the basic molecules of life itself — amino acids — exist in space.

Kant (in 1755) and Laplace (in 1796) first advanced the nebular hypothesis of the formation of the solar system. According to this, the sun and planets were formed out of the condensation of an immense cloud of matter. This seemed to fit the facts, and, by the time Engels wrote The Dialectics of Nature, it was generally accepted. In 1905, however, Chamberlain and Moulton, put forward an alternative theory — the planetesimal hypothesis. This was further developed by Jeans and Jeffreys, who advanced the tidal hypothesis in 1918. This involved the idea that the solar system originated as a result of a collision of two stars. The problem with this theory is that, if it were true, planetary systems would be extremely rare phenomena. The vast distances separating stars mean that such collisions are 10,000 times less common than supernovae — themselves far from common occurrences. Once again, we see that, by attempting to solve a problem by resorting to an accidental external source like a stray star, we create more problems than we solve.

Eventually, the theory which was supposed to have displaced the Kant-Laplace model was shown to be mathematically unsound. Other attempts, like the "three-star collision" (Littleton) and Hoyle's supernova theory, were also ruled out in 1939, when it was proved that the material drawn from the sun in such a way would be too hot to condense into planets. It would merely expand into a thin gas. Thus, the catastrophe-planetesimal theory was overthrown. The nebular hypothesis has been reinstated, but on a higher level than before. It is not merely a repetition of the ideas of Kant and Laplace. For instance, it is now understood that the clouds of dust and gas envisaged in the model would have to be much bigger than they thought. On such huge scales, the cloud would experience turbulence, creating vast eddies, which would then condense into separate systems. This perfectly dialectical model was developed in 1944 by the German astronomer Carl F. von Weizsäcker, and perfected by the Swedish astrophysicist, Hannes Alfvén.

Weizsäcker calculated that there would be sufficient matter in the largest eddies to create galaxies in the process of a turbulent contraction, giving rise to sub-eddies. Each of these could produce solar systems and planets. Hannes Alfvén made a special study of the magnetic field of the sun. In the early stages, the sun was spinning at a great speed, but was eventually slowed down by its magnetic field. This passed on angular momentum to the planets. The new version of the Kant-Laplace theory, as developed by Alfvén and Weizsäcker, is now generally accepted as the most likely version of the origins of the solar system.

The birth of our solar system some 4.6 billion years ago developed out of a cloud of shattered debris of a now extinct star. The present sun coalesced at the center of the revolving flat cloud, whereas the planets developed at different points encircling the sun. It is believed that the outer planets — Jupiter, Saturn, Uranus and Pluto — are a sample of the original cloud: hydrogen, helium, methane, ammonia and water. The smaller inner planets — Mercury, Venus, Earth and Mars — are richer in heavier elements and poorer in gases like helium and neon, which were able to escape their weaker gravities.

Aristotle thought that everything on earth was perishable, but that the heavens themselves were changeless and immortal. Now we know differently. As we gaze with wonder at the immensity of the night sky, we know that every one of these heavenly bodies which light up the darkness will one day be extinguished. Not only mortal men and women, but the stars themselves that bear the names of Gods experience the agony and the ecstasy of change, birth and death. And, in some strange way, this knowledge brings us closer to the great universe of nature, from which we came and to which we must one day return. Our sun has at present enough hydrogen to last for billions of years in its present state. Eventually, however, it will increase its temperature to the point where life on earth will become impossible. All individual beings must perish, but the wonderful diversity of the material universe in all its myriad manifestations is eternal and indestructible. Life arises, passes away, and arises again and again. Thus it has been. Thus it will ever be.

The birth and death of stars constitute a further example of the dialectical workings of nature. Before it runs out of nuclear fuel, the star experiences a prolonged period of peaceful evolution lasting millions of years. But on reaching the critical point, it experiences a violent end, collapsing under its own weight in less than a second. In the process, it gives off a colossal amount of energy in the form of light, emitting more in a few months than the sun emits in a billion years. Yet this light represents only a small fraction of the total energy of a supernova. The kinetic energy of the explosion is ten times greater. Perhaps ten times more than the latter is carried away in the form of neutrinos, emitted in a split-second flash. Most of the star's mass is scattered into space. Such a supernova explosion in the vicinity of the Milky Way hurled forth its mass, reduced to nuclear ashes, containing a large variety of elements. The earth and all that is in it, ourselves included, is entirely composed of this recycled stardust, the iron in our blood being a typical sample of recycled cosmic debris.

These cosmic revolutions, like the earthly variety, are rare events. In our own galaxy, only three supernovas have been recorded over the past 1000 years. The brightest of these, noted by Chinese observers in 1054, produced the Crab Nebula. Moreover, the classification of stars has led to the conclusion that there is no new kind of matter in the universe. The same matter exists everywhere. The main features of the spectra of all stars can be accounted for in terms of substances that exist on earth. The development of infrared astronomy provided the means of exploring the interior of dark interstellar clouds, which are probably where most new stars are formed. Radio astronomy has begun to reveal the composition of these clouds — mainly hydrogen and dust, but with an admixture of some surprisingly complex molecules, many of them organic.

NOTES PART TWO

1 B. Hoffmann, *The Strange Story of the Quantum*, p. 147.

2 Engels, *Dialectics of Nature*, p. 92.

3 B. Hoffmann, op. cit., pp. 194-5.

4 *Financial Times*, 1/4/94, our emphasis.

5 I. Asimov, *New Guide to Science*, p. 375.

6 D. Bohm, *Causality and Chance in Modern Physics*, pp. 86 and 87.

7 T. Ferris, *The World Treasury of Physics, Astronomy, and Mathematics*, pp. 103 and 106.

8 E. J. Lerner, *The Big Bang Never Happened*, pp. 362-3.

9 *LCW*, Vol. 14, p. 55.

10 T. Ferris, op. cit., p. 95-6.

11 Spinoza, *Ethics*, p. 8.

12 Quoted in I. Stewart, *Does God Play Dice?* pp. 10-2.

13 Engels, *The Dialectics of Nature*, pp. 289-90.

14 D. Bohm, op. cit., p. 20.

15 J. Gleick, *Chaos, Making a New Science*, p. 124.

16 D. Bohm, op. cit., pp. x and xi.

17 Ibid., pp. 50-1.

18 B. Hoffmann, op. cit., p. 152.

19 D. Bohm, op. cit., pp. 25 and 4.

20 Hegel, *Philosophy of Right*, p. 10.

21 *MESW*, Vol. 3, pp. 338-9.

22 *MESC*, Marx to Kigelmann, 17th April 1871, p. 264.

23 Ibid., *Engels to Starkenburg, 25th January 1894*, p. 467.

24 Engels, *The Dialectics of Nature*, pp 17 and 304.

25 Engels, *Anti-Dühring*, p. 32.

26 Quoted in Gleick, op. cit., pp. 251-2.

27 *Marx and Engels, Collected Works*, Vol. 4, p. 93, henceforth referred to as *MECW*.

28 Freud, *The Psychopathology of Everyday Life*, p. 193.

29 Job 14: 1

30 Aristotle, op cit., pp. 342 and 1b.

31 Hegel, *Phenomenology of Mind*, p. 151.

32 Prigogine and Stengers, op. cit., p. 89.

33 Hegel, *Phenomenology of Mind*, p. 104.

34 Hegel, *Science of Logic*, Vol. 1, p. 229.

35 Landau and Rumer, *What is Relativity?* pp. 36 and 37.

36 Feynman, op. cit., Vol. 1, 1-2.

37 Trotsky, *The Struggle Against Fascism in Germany*, p. 399.

38 Feynman, op. cit., chapter 5, p. 2.

39 N. Calder, *Einstein's Universe*, p. 22.

40 J. D Bernal, *Science in History*, pp. 527-8.

41 N. Calder, op. cit., p. 13.

42 Asimov, op. cit., p. 359.

43 Hegel, *The Phenomenology of Mind*, p. 151.

44 K. Popper, *Unended Quest*, pp. 96-7 and 98.

45 Prigogine and Stengers, op. cit., pp. 10, 252-3 and 277.

46 Quoted in Lerner, op. cit., p. 134.

47 P. Davies, *The Last Three Minutes*, pp. 98-9.

48 Quoted by Davies, op. cit., p. 13.

49 M. Waldrop, *Complexity*, pp. 33-4.
50 J. Gleick, op. cit., p. 308.
51 Lerner, op. cit., p. 139.
52 Prigogine and Stengers, op. cit., p. 298.
53 Ibid., pp. 148, 206 and 287.
54 Quoted by Lerner, op. cit., p. 214.
55 Ibid., p. 152.
56 Ibid., p. 158.
57 Ibid., pp. 39-40.
58 *The Rubber Universe*, pp. 11 and 14, our emphasis.
59 Quoted in Lerner, op. cit., pp. 164-5.
60 P. Davies, op. cit., pp. 123, 124-5 and 126.
61 Lerner, op. cit., p. 14
62 Ibid., pp. 52, 196, 209 and 217-8.
63 Ibid., pp. 153-4, 221 and 222.
64 Ibid., p. 149.
65 T. Ferris, op. cit., p. 204.
66 S. W. Hawking, *A Brief History of Time, From the Big Bang to Black Holes*, p. 34.
67 Ibid., pp. 46-7 and 33.
68 Engels, *Anti-Dühring*, pp. 64-5.
69 Ibid., p. 68.
70 Hawking, op. cit., pp. 50 and 88-9.
71 Ibid., p. 89.
72 Engels, *The Dialectics of Nature*, pp. 68-9.
73 Hawking, op. cit., p. 116.
74 Engels, *Anti-Dühring*, pp. 62-3.
75 Lerner, op. cit., p. 161.
76 W. Rees-Mogg and J. Davidson, op. cit., p. 447

APPENDIX I

WHAT THE HUMAN GENOME MEANS FOR SOCIALISTS

An Epoch-making Discovery

Once every century or so great scientific breakthroughs grip the imagination of the world. With the publication of the results of the human genome project, we stand on the threshold of such a breakthrough. Science is now poised to understand the forces behind evolution, explode racial myths, change the way doctors diagnose disease, and try to help people live longer. The new approach — looking at systems of genes rather than individual genes — will transform biologists' view of the human body. It is approximately the equivalent of Mendeleyev's periodic table in chemistry, or the breakthrough made by Watson and Crick 48 years ago, when they first described the double helix shape of DNA. "Before we were looking through a keyhole," says James Pierce, a professor of genetics at the University of the Sciences in Philadelphia. "Now, the door is open."

The scope of this project was immense. Roughly 2,000 scientists world-wide took part in the sequencing effort. The research project was actually carried out by two different groups — one funded by the US government, the other by the UK based Wellcome Trust's Glaxo Sanger Center. Both groups have come to the same wholly unexpected conclusion, which is that the number of genes in the human genome is less than one quarter of the anticipated result. One team, led by J. Craig Venter of Celera Genomics Corp., found strong evidence of 26,383 genes and weaker evidence for an additional 12,731. The other team said there are probably about 35,000 genes, but possibly as many as 40,000. The Celera team published its findings in the journal Science; the international team published in Nature. "It is good to have a rough agreement between the two sides," said Venter. "Certainly, it shows that there are far fewer genes than anyone imagined." Researchers also reported that each human gene can make two proteins or more, upsetting the long-held notion that one gene makes one protein.

These discoveries have profound implications for medical and drug research. They suggest that genes play less of a role in causing disease, and many of our other traits, than many researchers realized. And new approaches to treatment will not only have to focus on genes and how they function but also on how they interact. Even in the short term the practical implications are immense. Advances in medical science could follow from the identification of the genes responsible for non-inherited diseases, and scientists could eventually learn to predict the likelihood of someone developing a genetic disorder. Medicine could then tailor specific drugs for patients and there could be a gradual increase in the range of treatments for genetic diseases in the next few decades. Within five to seven years we may see this work bearing fruit in areas like diabetes, heart disease and major mental disorders. Major advances had already been made in schizophrenia.

The discoveries made by the Human Genome Project have dramatically confirmed the position of Marxism, as expounded in *Reason in Revolt* six years ago. For decades, a large number of geneticists have argued that everything from intelligence to homosexuality and criminality was determined by our genes. The most reactionary conclusions have been drawn from these assumptions: for example, that black people and women are genetically conditioned to be less intelligent than white people and men; that rape and murder are somehow natural, because they are genetically determined; that there is no point in spending money on schools and houses for the poor because their poverty is rooted in genetics and therefore cannot be remedied. Above all, that the existence of inequality is natural and inevitable, and that all attempts to abolish class society are futile, since it is somehow rooted in our genes. This was a very good example of how science cannot be separated from politics and class interests, and how the most eminent scientists can be pressed — consciously or not — into the services of reaction. But leaving aside for a moment the social and political implications, in purely scientific terms this is a defining moment in history.

The Puzzle of the Missing Genes

Despite the enormity of their discovery, the biologists who reported their first analysis of the decoded sequence were clearly as perplexed as they were enlightened. The chief puzzle is the unexpectedly small number of human genes. When they first screened the gene families likely to have new members of interest to pharmaceutical companies, "there was almost panic because the genes weren't there," Dr. Venter said. The problem is that the textbooks have long estimated the number of human genes to be far greater. The string of biological code present in humans was so long — some 3 billion units — that scientists had expected it to contain instructions to create anywhere from 50,000 to 150,000 genes.

This assumption was based on a comparison with simpler organisms such as fruit flies. It was argued that, if the humble fruit fly had 13,000 genes, then a human being — a far bigger and more complex entity — must have many times more. The estimate of up to 150,000 genes seemed reasonable after the first two animal genomes were deciphered. The laboratory roundworm, sequenced in

December 1998, has 19,098 genes and the fruit fly, decoded last March, has 13,601 genes. Dr. Randy Scott, chief scientific officer of Incyte Genomics, predicted in September 1999 that there were 142,634 human genes. But the human gene complement has now turned out to be far closer to genetic patrimony of these two tiny invertebrates than almost anyone had expected. Instead, they have discovered that vast stretches of the code create very few genes. "We have about twice as many genes as a fly and the same number as corn," Venter says. "Think of that the next time you eat corn." Last week Dr. Scott said he accepted the rationale for the lesser number and now puts the human complement at around 40,000.

Celera's rival, the publicly funded consortium of academic centers, has come to a similar conclusion. Its report in this week's Nature puts the probable number of human genes at 30,000 to 40,000. Because the current gene-finding methods tend to over predict, each side prefers the lower end of its range, and 30,000 seems to be the more accurate estimate. The two teams found other contradictions, too. Most of the repetitive DNA sequences in the 75 percent of the genome that is essentially "junk" ceased to accumulate millions of years ago, but a few of sequences are still active and may do some good. The chromosomes themselves have a rich archaeology. Large blocks of genes seem to have been extensively copied from one human chromosome to another, a fact which will encourage genetic archaeologists to work out the order in which the copying occurred and thus to reconstruct the history of the animal genome.

The small number of human genes poses a dilemma for scientists. As the modest number of human genes became apparent, biologists in both teams were forced to think how to account for the greater complexity of people, given that they seem to possess only 50 percent more genes than the roundworm. If humankind only has 13,000 more genes than *Caenorhabditis elegans* (a roundworm) or 6,000 more than *Arabidopsis thaliana* (a weed), what makes people so advanced by comparison? To quantify the position: the roundworm is a little tube of a creature with a body of 959 cells, of which 302 are neurons in what passes for its brain. Humans have 100 trillion cells in their body, including 100 billion brain cells. Despite the fashionable tendency to deny the existence of progress in evolution, it is surely reasonable to suppose there is something more to *Homo sapiens* than a roundworm like *Caenorhabditis elegans*!

The Christian Science Monitor posed the question thus: "If man is so advanced, how come his gene count doesn't look that much different from a weed's or a worm's?" And if, as suspected, the chimpanzee genome turns out to be very similar to the human genome, then scientists will still have to explain how one species has come to so dominate the world in the past 50,000 to 150,000 years while others are still climbing trees. This question, however, cannot be answered purely in terms of genetics. The great advantage of the recent discoveries is that they have moved away from the concept that everything could be explained in terms of individual genes. The human genome can now be approached as a complex totality. Genes have to be understood, not as a collection of entities but as a process of highly complex interactions. The further exploration of these interactions, their history and the resulting genetic "archeology" will eventually give us a true understanding of ourselves and our place in the nature of things. There can be no more important subject for human beings.

Biological Determinism Exposed

Marxists have, of course, never ignored the role of genetics in shaping human behavior. It goes without saying that genes play a most important role. They provide to some extent the raw material out of which individual humans are developed. But they represent only one side of a very complex equation. The problem arises when certain people attempt to present genes as the sole agent conditioning human development and behavior, as has been the case for quite some time now. In reality, genes ("nature") and environmental factors ("nurture") interact upon each other, and that in this process, the role of the environment, which has been systematically denied or downplayed by the biological determinists, is absolutely crucial.

The recent revelations of the human genome project have decisively settled the old "nature" — "nurture" controversy. The relatively small number of genes rules out the possibility of individual genes controlling and shaping behavior patterns such as criminality and sexual preference. It completely destroys the case of people like Dean Hammer who claimed to have isolated a gene on the human X-chromosome which allegedly disposes people to homosexuality. Similar claims have been made for a whole series of human traits from running ability to artistic taste and even political tendencies! In reality human behavior is extremely complex and cannot be reduced to genetics. The latest discoveries flatly contradict all the nonsense which has been put forward for years as irrefutable.

The biological determinists insisted that in some way genes are responsible for things, like homosexuality and criminality. They attempted to reduce all social problems to the level of genetics. In February 1995, a conference on Genetics of Criminal and Anti-Social Behavior was held in London. Ten of the thirteen speakers were from the United States where a similar conference in 1992 with racist overtones was abandoned because of public pressure. While the chairperson, Sir Michael Rutter of the London Institute of Psychiatry stated "there can be no such thing as a gene for crime," other participants, like Dr. Gregory Carey of the Institute of Behavioral Genetics, University of Colorado, maintained that genetic factors as a whole were responsible for 40-50% of criminal violence. Although he said it would be impractical to "treat" criminality through genetic engineering, others said there were good prospects for developing drugs to control excessive aggression, once the responsible genes had been found. He suggested, however, that abortion should be considered when antenatal testing indicates a child is likely to be born with genes predisposing it to aggression or antisocial behavior. His view was endorsed by Dr. David Goldman from the Laboratory of Neurogenetics at the US National Institutes of Health. "The families should be given the information and should be allowed to decide privately how to use it" (*The Independent*, 14th February 1995).

There are many other examples. The notorious *Bell Curve* by Charles Murray, resurrected the old argument that genetics explains the gap between the average IQ of American whites and blacks. C. R. Jeffery wrote that "Science must tell us what individuals will or will not become criminals, what individuals will or will not become victims, and what law enforcement strategies will or will not work." Yudofsky reinforces Jeffery's enthusiasm with his assertion: "We are now on the verge of a revolution in genetic medicine. The future will be to understand the genetics of aggressive disorders and to identify those who have greater tendencies to become violent."

When we criticized these false theories in *Reason in Revolt*, we had no means of knowing that in a few years their unscientific character would be so clearly demonstrated. Now the revelation that the number of genes in humans is not more than 40,000 and possibly as few as 30,000 or less has shattered the case for biological-genetic determinism at a single stroke. Dr. Craig Venter, the US geneticist whose company Celera was one of the main groups responsible for the sequencing project, put the matter very simply: "We simply do not have enough genes for this idea of biological determinism to be right. The wonderful diversity of the human species is not hard-wired in our genetic code. *Our environments are critical*" (Observer, 11/2/2001, my emphasis).

The Observer goes on to explain:

> It is only when scientists looked at the way these genes are switched off and on and made to manufacture proteins that they could see a significant difference between various mammalian species. The key difference lies in the manner in which human genes are regulated in response to environmental stimulation compared with other animals.

That is to say, it is the environment — the external stimuli of both the physical world and the conditions in which we live — that condition evolution in a decisive way. The role of genes is important, but the relation between genes and development is not simple and mechanical, as maintained by the crude theory of biological determinism, but *complex and dialectical,* as argued by Marxism. Let us take one example of the dialectical interaction between genes and environment: perfect pitch. In his new book *The Sequence,* which describes the search for the human genome, Kevin Davies writes:

> There has been a recent study on perfect pitch, the ability to know the absolute pitch of a musical note, that strongly suggests that it is acquired through the inheritance of a single gene.

> This may sound like a clear-cut case of biological determinism. However, there is a crucial corollary — you have to be exposed to early music training for the ability to materialize. In other words, even in seemingly simple inherited abilities, nurture has a role to play.

Thus, there is a complex interplay between the genetic composition of the organism and the physical conditions that surround it. In Hegelian language, the genes represent *potential.* But this potential is only activated by external stimuli. The genes are "switched on" by the environment, producing small changes, some of which prove to be useful from an evolutionary point of view, although in fact most genetic mutations are harmful or confer no benefit. Over a period, the beneficial mutations give rise to qualitative changes in the organism, giving rise to the process we refer to as natural selection.

The Observer's Editorial drew the political conclusions:

"Politically, it offers comfort to the Left, with its belief in the potential of all, however deprived their background. But it is damning to the Right, with its fondness for ruling classes and original sin" (*Observer* Editorial, 11/ 2/ 2001, my emphasis).

"Race Has No Meaning in Science"

The results of these investigations are highly significant from another point of view. The genome reveals the existence of unity in human diversity. They completely destroy the myth of racial superiority. The biological essence of human populations is the same. The absence of a race gene was confirmed from two different directions. Celera used DNA from males and females who described themselves as Asian Chinese, African American, Caucasian and Hispanic Mexican. Scientists could not distinguish one ethnicity from another. *No gene by itself or together with others could predict the race of those studied.*

The new research suggests that all individuals are 99.99 percent alike. And researchers are finding that the gene pool in Africa, where humankind is thought to have originated, remains more diverse than in the rest of the world. These findings completely undermine all notions of differences based on skin color. Svante Pääbo, a German researcher, noted in an essay published in *Science* magazine with the release of the draft genome sequence explains: "It is often the case that two persons who descend from the same part of the world, and look superficially alike, are less related to each other than to persons from other parts of the world who may look very different."

Dr. Eric Lander of the Whitehead Institute for Genome Research, and part of the international consortium, pointed out the fact that any two people were 99.9 per cent genetically identical still left room for considerable genetic variation. One tenth of one per cent of human genes account for hereditary differences. Basically, all human beings are the same. The research on the human genome has proved beyond doubt that while outwardly we may be different, genetically we are 99% identical. Only about 3 million of the 3 billion chemicals in the genome differ from one person to another, which makes distinctions such as race scientifically meaningless. Ethnic and cultural differences among different groups of humans undoubtedly exist, but these differences are insignificant at the genetic level where people are remarkably the same, regardless of race and gender. Racial hatred cannot therefore be justified and rationalized as arising from genetic differences.

The Seattle Times Editorial of February 13, points out:

> One bonus of the human genome project is to knock down bigots who have long strained to camouflage old-fashioned hatred with scientific prattle about genetic superiority. The road map of the human DNA sequence leads to one conclusion: *Race has no meaning in science* (my emphasis).

This will, of course, not put an end to racism, which is rooted in the contra-dictions of capitalism in the epoch of its decline. But at least it will rob the purveyors of racist poison of the fig-leaf of pseudo-scientific arguments. In future, any attempt of racists and bigots to appeal to science to support their views will be met with the contempt it deserves.

"If anything, such studies will have the opposite effect because prejudice, oppression and racism feed on ignorance," Dr. Pääbo writes. Pääbo argues knowledge of the genome should foster compassion: "Consequently, stigma-tizing any particular group of individuals on the basis of ethnicity or carrier status for certain (genes) will be revealed as absurd."

Creationism Exploded

The revelation of the genome's long and complex history, so long hidden from view, has prompted discussions about the nature of man and the process of creation. Incredibly, in the first decade of the twenty first century, the ideas of Darwin are being challenged by the so-called Creationist movement in the USA, which wants American schoolchildren to be taught that God created the world in six days, that man was created from dust and that the first woman was made out of one of his ribs, the Almighty presumably being on an economy drive that day.

The Creationist movement is no joke. It involves millions of people and is —incredibly — spearheaded by scientists, including some geneticists. This is a graphic expression of the intellectual consequences of the decay of capitalism. It is an extremely striking example of the dialectical contradiction of the lag of human consciousness. In the most technologically advanced country in the world, the minds of millions of men and women remain under the influence of the accumulated rubbish of the past. In the same way that we share our genes with the most primitive organisms, and a great many of them are "junk genes" that play no useful role but remain as a residue of the prehistory of our species, so we find stored in the deep recesses of human consciousness relics of the most primitive superstitions and prejudices. These are remnants of a barbarous and half-forgotten past that has vanished but is not yet overcome. In the consciousness of the Creationists, and especially their leading spokesmen and women we find echoes of the times when men sacrificed prisoners of war to the gods, prostrated themselves before graven idols and burnt witches at the stake. If this movement were to succeed, as one scientist recently put it, we would be back in the Dark Ages.

The latest discoveries have finally exploded the nonsense of Creationism. It has comprehensively demolished the notion that every species was created separately, and that Man, with his eternal soul, was especially created to sing the praises of the Lord. It is now clearly proved that humans are not at all unique creations. The results of the human genome project show conclusively that we share our genes with other species — that ancient genes helped to make us who we are. Humans share their genes with other species going far back into the mists of time. In fact, a small part of this common genetic inheritance can be traced back to primitive organisms such as bacteria. "Evolution no longer has time to make new genes. It must make new genes from old parts," observes Eric Lander of the Whitehead Institute for Genomic Research in Cambridge, Mass. The two teams found an astonishing degree of gene conservation over the past 600 million years of evolution on earth: "In many cases we have found that humans have exactly the same genes as rats, mice, cats, dogs and even fruit flies," Venter continues. "Take the gene PAX-6. We have found that when it is damaged, eyes will not form. You can take a human gene, insert it into fruit flies, and the vision of their offspring will be restored."

Scientists have now found some 200 genes that humans share with bacteria — a revelation which surprised James Watson, the discoverer of DNA and arguably the world's most renowned geneticist: "We knew that genes jumped between bacteria," he commented, "but not that they jumped between bacteria and man." In this way, the final proof of evolution has been established. In a fundamental way, these genetic "fossils" have helped over billions of years of evolution to make us what we are. "No doubt the genomic view of our place in nature will be both a source of humility and a blow to the idea of human uniqueness," Svante Pääbo writes in a separate article in *Science* magazine. And, he continues, "the realization that one or a few genetic accidents made human history possible will provide us with a whole new set of philosophical challenges to think about." For Marxists too, the human genome holds important philosophical implications.

Science and Dialectics

At the London launch of research to decipher the genetic code, Sir John Sulston, former director of the Sanger Center, described the drawing of the human gene map as "a remarkable, iconic event in the era of molecular biology:

"It is remarkable that a living organism has got so smart and has made such clever machines that it can think about what it is doing, that it has actually read out the code, the instructions, to make itself. It's the sort of thing that causes philosophers to disappear up their navels if they think too hard about it. It really is a superficial paradox . . . but it's true. We are understanding how we work," he was quoted as saying.

Indeed, the spectacular march of science in our epoch makes the speculations of philosophy seem pale and uninteresting by comparison. The deeds of humanity have by far outstripped the general level of its consciousness, which remains largely mired in the barbarous past. The new discoveries provide the human race with inspiration and confidence in itself. It provides us with a vision of ourselves, who we really are, and where we have come from — perhaps also where we are going to.

Nevertheless, despite Sir John Sulston's disparaging remarks about philosophy, there are still a few areas where a knowledge of real philosophy would undoubtedly benefit scientists. Of course, there is philosophy and philosophy! Very little of what passes for philosophy in the universities nowadays is of any use to scientists — or anyone else. But there is one honorable exception, which is still awaiting the recognition that is long overdue: that is, dialectical materialism. Although many of the main tenets of dialectical materialism have re-surfaced in recent years, incorporated into the theories of chaos, complexity and, more recently, ubiquity, this debt has never been acknowledged. Dialectics in science is, to paraphrase Oscar Wilde, the philosophy that dares not speak its name. This is a pity, since a knowledge of the dialectical method would certainly have helped avoid a number of pitfalls into which science has occasionally strayed as a result of incorrect assumptions. The human genome is a case in point.

Of course, there is no question of any philosophy dictating to science. The results of science must be determined by its own methods of investigation, observation and experiment. Nonetheless, it is a mistake to imagine that scientists approach their subject matter without any philosophical assumptions. Behind every hypothesis there are always many assumptions, not all of them derived from science itself. The role of formal logic, for example, is taken for granted. It is an important role, but one that has definite limitations. Trotsky explained that the relationship between formal logic and dialectics resembles that between elementary mathematics and calculus. The great advantage of dialectics over formal logic is that it deals with things in their motion and development, and moreover shows how all development takes place through contradictions. Thus, Marx predicted that the line of evolution was not a straight line, but a line in which long periods of slow development ("stasis" in modern terminology) was broken by sudden leaps — breaks in continuity that impelled the process in a new direction.

Let us take one instance. The dialectical method explains how quite small changes can, at a critical point, produce enormous transformations. This is the famous law of the transformation of quantity into quality — a wonderful and all-embracing law — that was first worked out by the ancient Greeks, and later fully developed by Hegel and placed on a scientific (materialist) basis by Marx and Engels. The importance of this law has only recently been recognized by science through chaos theory. The latest version of this ("ubiquity") has demonstrated that this law has a universal character and is of key importance in many of the most fundamental processes in nature. It has a crucial bearing on the present discussion.

What is the source of the error which led geneticists to conclude that humans possessed far more genes than is, in fact, the case? It is known in philosophy as *reductionism*, and flows from the mechanical assumption that nature knows only purely quantitative relations. This lies at the heart of biological determinism which approaches humans as a collection of genes, and not as complex organisms, processes, the product of a dialectical interrelation between genes and the environment. Their mode of reasoning is that of formal logic, not dialectics. And from this philosophical standpoint, their conclusions were quite consistent. Logical — but radically false. They reasoned that, since humans are bigger and more complex than fruit worms and roundworms, they must have vastly more genes. However, nature produces many examples to show how changes in quantity eventually beget changes in quality. In many instances, quite small modifications can produce huge changes. The apparent contradiction between the size and complexity of humans and the relatively small number of genes involved can only be explained by recourse to this law.

In *Reason in Revolt*, we subjected this method to a comprehensive criticism. Dealing with the method of Richard Dawkins in *The Selfish Gene*, we wrote:

> Dawkins' method leads him into the swamp of idealism, when he attempts to argue that human culture can be reduced to units he calls memes, which, apparently, like genes, are self-replicating and compete for survival. This is clearly wrong. Human culture is passed down from generation to generation, not through memes, but through education in the broadest sense. It is not biologically inherited but has to be painstakingly relearned and developed by each new generation. Cultural diversity is bound up not with genes but social history. Dawkins' approach is essentially reductionist.

In a commentary in Science magazine, Dr. Jean-Michel Claverie, of the French National Research Center in Marseilles, notes that with a simple combinatorial scheme, a 30,000-gene organism like the human can in principle be made almost infinitely more complicated. This is a perfect example of the transformation of quantity into quality. Dr. Claverie suspects humans are not that much more elaborate than some of their creations. "In fact," he writes, "with 30,000 genes, each directly interacting with four or five others on average, the human genome is not significantly more complex than a modern jet airplane, which contains more than 200,000 unique parts, each of them interacting with three or four others on average."

The initial scanning of the genome suggests two specific ways in which humans have become more complex than worms. One comes from analysis of what are called protein domains. Proteins, the working parts of the cell, are often multipurpose tools, with each role being performed by a different section or domain of the protein. Many protein domains are very ancient. Comparing the domains of proteins made by the roundworm, the fruit fly and people, the consortium reports that only 7 percent of the protein domains found in people were absent from worm and fly, suggesting that "few new protein domains have been invented in the vertebrate lineage."

The most important thing to grasp is that very small genetic mutations can give rise to huge differences. For example, the genetic difference between humans and chimpanzees is less than two percent. As the latest research shows, we have a lot more in common with other animals than we would perhaps like to admit! Most of the genetic material present in modern humans is very old, and identical with the genes which are found even in such lowly beings as fruit flies. Nature is inherently conservative and economical in its workings! Organic matter has evolved from inorganic matter, and higher life forms have evolved from lower ones. We share most of our genes, not just with monkeys and dogs, but with fishes and fruit flies. But merely to state this fact is insufficient. It is also necessary to explain the dialectical process whereby one species is transformed into another. It has recently become fashionable to blur the difference between humans and other animals, in what is obviously an over-reaction against the old idea of Man as a special creation, placed by the Almighty over all Creation.

It has become fashionable to deny the existence of any progress at all, presumably in the interests of an ill-conceived evolutionary "democracy." As a matter of fact, the genetic difference between humans and chimpanzees may be less than two percent, but what a difference that makes! It is a dialectical leap that transforms quantity into quality. But unfortunately, dialectics is subjected to a conspiracy of silence in the universities and consequently remains completely unknown to most scientists. The most likely explanation for how to generate extra complexity other than by adding more genes is the idea of *combinatorial complexity* — that is, with just a few extra proteins one could make a much larger number of different combinations between them to produce *a qualitative change.* The matter has not yet been decisively settled, and much more research will be necessary. But there is little doubt that the final solution will be found somewhere along these lines.

The Human Genome and Big Business

Scientists from the Human Genome Project described the mapping of the human genetic code as "a gift to the world" that could improve the ability to detect disease and encourage the development of new medicines. That is undoubtedly what it ought to be. But in the market economy, such gifts invariably carry a hefty price tag.

Mapping of the human genome is a historic achievement, but the task of clarifying the complex and dialectical process by which genes interact with environmental factors has only begun. Science has yet to discover the role of genes in complex diseases The possibilities are limitless, but this immense potential for human progress immediately comes into collision with the narrow limits of the capitalist system where everything is subordinate to private gain. The new technology will be monopolized by the big multinational companies who will exploit it for themselves. The general interests of humanity will come a poor second. Already the issues of privacy, social and legal impacts, regulation and ethical research are generating heated controversy.

The human genome project has naturally attracted the attention of Big Business, scenting the prospect of big profits. The result, however, has caused consternation in the board rooms of some of the big pharmaceutical companies that were looking forward to using it to make money out of the new cures that will hopefully emerge from a deeper knowledge of the behavior of genes. Initially it was assumed that there were about 120,000 to 150,000 genes involved. The drugs companies made their investments accordingly. So when Craig Venter and his team issued a preliminary report, indicating that the number of genes would be "only" about 80,000, he received an angry phone call from the head of a leading biotechnological company:

"He was cursing and swearing and using all sorts of obscenities about my company and about myself." When asked by Venter to state his problem, the unnamed chief replied: "You've just announced there are only 80,000 human genes and I've just done a deal with SmithKline-Beecham. I've agreed to sell them 100,000 genes — where do you suppose I am supposed to get the rest, you bastard." It is perhaps just as well that this capitalist died before he learned that the real figure was not 80,000 but around 30,000!

This little exchange casts an amusing light on the relation between Big Business and scientific research. Scientists   at least the good ones — are interested in pursuing knowledge for its own sake, of breaking new ground and pushing forward the horizons of science. Big Business is interested only in making money. In this case, they have been prepared to invest because they see the prospects for juicy profits. The biotechnological industry is based on isolating genes that go wrong in our bodies in order to create new drugs which they can sell for a profit. Even 30,000 potential new drugs spells a lot of money — for some.

The International Human Genome Sequencing Consortium is a multinational, publicly financed project that makes its findings available to all. But Celera Genomics, a private, for-profit venture, is keeping its findings closely held, hoping to make its investors wealthy. With its map of the human genome in its pocket, the Celera Genomics Group is hoping to cash in as drug and biotech companies pay to look at the genetic information that could help them develop new medicines and treatments. Although the Human Genome Project offers the mapping and readers' guides free of charge, research firms such as Immunex have already been using the genome database, paying a reported US$15 million to get their gene mapping from Celera. Analysts say the company, which has a market capitalization of around $3 billion, is positioned to move from a genetic librarian role to developing drugs and treatments that will be made easier by the genome map. Shares of Celera were trading up 10 cents to $47.85 in trading on the New York Stock Exchange on Tuesday 13th, after rising 15 percent the previous day. Already Celera is reportedly using the genome mapping to develop drugs and treatments on its own. It is said that Celera may become as big as drug and health giant Pfizer.

Because companies usually want to have secure ownership rights to genes before investing the millions of dollars it takes to develop drugs from them, doubts about patent rights could have far-reaching effects. Some researchers have said this suggests that two scientists or companies, while researching different proteins involved in different diseases, are likely to have sought to patent portions of the same gene. The result could be a series of clashes over patents that would block one or both companies from continuing their research, producing a drug or developing a genetic test for disease. "I think there are lots of suits to be filed, and this [the low number of genes] will make it more so," said Dr. Robert H. Waterston, a DNA-mapping specialist at Washington University in St. Louis.

Many others agree. "I think it could inhibit research," said Arthur Caplan, a bioethics professor at the University of Pennsylvania and an advisor to Celera. "The question that will face companies is: What does corporate responsibility require that you do in terms of sharing access, making the information you own available? They're going to have a stewardship responsibility, because they work in health. This is not like patenting Coke." If a flare-up of patent clashes "hasn't happened yet, it's going to happen soon," said Dr. Lee Hood, a molecular biologist at the University of Washington in Seattle. "I think there will be genes with 10 or 50 different [protein] forms. . . . You will have patents for every splicing and, how that gets untangled, God only knows."

The patent office estimates that it has issued patents on about 1,000 full-length human genes, but it has tens of thousands of applications pending. The vultures are already circling! The prospect for chaos and endless lawsuits is clear and will work to the detriment of science and, ultimately, the millions of people who are desperately in need of new medical treatments, made possible by the genome project. Even when two valid patents are issued, one owner may be able to win a lawsuit blocking another from moving ahead with research or with a drug. Critics have already complained that Human Genome Sciences had been granted patent rights for the gene's role in AIDS without sufficient proof that it understood that role. This is only the beginning.

There are other problems concerning the use of this technology under capitalism. It could usher in a new era of genetic discrimination. For example, if scientists create diagnostic tests that can determine an individual's predisposition to certain diseases, should that person's insurance company or employer know about it? "Without adequate safeguards, the genetic revolution could mean one step forward for science and two steps backwards for civil rights," write United States Senators James Jeffords and Tom Daschle in an article in *Science* magazine. "Misuse of genetic information could create a new underclass: the genetically less fortunate." Dr. Venter and Dr. Collins, the leading pioneers in this field, have both deplored attempts by companies to test workers secretly and discriminate against them on the basis of their genetic profiles. Recently, for the first time ever, the federal Equal Employment Opportunity Commission sued an employer — Burlington Northern Santa Fe Railway — for discrimination based on genetic testing. In a survey of 2,133 employers last year by the American Management Association, seven said they are currently using genetic testing for job applicants or employees, according to Science magazine.

Like genetically modified food, or any other technological discovery, the human genome in the hands of greedy and irresponsible capitalists can be changed from a blessing on humanity to a curse. The latest marvelous discoveries in genetics, which were only made possible by the collaboration of men and women from every continent and nationality, and which go to the heart of that most profound question: who we are, cannot be monopolized by a handful of profiteers. The Labor Movement everywhere must demand the nationalization of the big bio and pharmaceutical companies, as the first step to nationalizing all the big banks and monopolies that dominate our lives and subject every aspect of our existence to the dictatorship of Capital. Only in a rationally planned socialist economy can the new discoveries achieve their full potential and be placed where they belong — at the service of humanity.

Limitless Possibilities

The mapping of the human genome has carried us one step nearer to the goal of developing our physical and intellectual capacities to the fullest extent. This process is as yet in its earliest infancy. The next great challenge is to understand how genes are regulated. Genes turn on and off in patterns. Understanding how this process works will be critical for developing new drugs for lingering diseases. That work is just beginning, but it promises to transform medicine. "In the 20th century we treated symptoms of diseases," Lander says. "In the 21st, we're going to treat the causes."

The tantalizing prospect opens up before us of a world free from the scourge of disease, the obliteration of cancer and AIDS — those modern equivalents of the Black Death, the eradication of malaria and all the other illnesses that spell misery, suffering and death for millions of the poorest people on earth. We have the realistic prospect of curing the mentally ill and the helpless victims of genetic disorders. All these are now practical propositions that can be realized within years or decades. But these things pale in insignif-icance before the longer term prospects that open up before us. In the long run, it is not inconceivable that human beings can attain mastery over the blind workings of natural selection itself. In the hands of private capitalists who put personal gain before all other considerations, genetic engineering poses a deadly threat even to the future of life on earth. But in a rationally ordered society, the new technology can pave the way to the most tremendous achievements yet seen. In the pages of the Bible, the blind saw, the deaf heard, the lame walked and the dead were raised. Now all these miracles can be achieved by science without recourse to the supernatural.

Of course, men and women will never achieve the kind of tedious immortality held out by religion. We should not desire to live forever, but to live this life — the only one available to us — to the full. The Bible promises us a life span of "threescore years and ten." Yet in the period of capitalism's senile decay, for countless millions, this is a dream. Life for the overwhelming majority of our planet in the first decade of the 21st century, in the celebrated words of Hobbes, remains nasty, brutish and short. Yet there is no reason why this should be the case. The potential of modern industry, agriculture, science and technique is more than enough to solve all the pressing needs of humanity and create a paradise for men and women, not in the cloudy realm of the Hereafter, but right here and now — a paradise in *this* world.

By making use of the benefits conferred upon us by science and technology, the ordinary human life span can be extended far beyond its present "natural limits." It s entirely possible to foresee a world in which it would be considered normal to live a healthy and active life beyond a hundred years: to live life to the full, to add to the total store of human achievements in art, science and all other fields of social activity, to raise ourselves up to the fullest potential permitted by Mother Nature, and then, when we have given all that we have to give, to retire from this world in good heart to make way for the new generations who will continue and extend our life's work. Such a perspective — essentially modest in the context of what we now know to be possible — could be considered "utopian" only by second-rate intellects and people who have become so demoralized and de-humanized by the decay of capitalism that they have lost all hope and all sense of human dignity, and have persuaded themselves that the present miserable state of affairs is all we can hope for.

What these wonderful achievements of science reveal to us is the limitless potential of the human race. And what it should also do is to make us all the more painfully aware of the criminal waste that is the most horrific feature of the so-called market economy. Until now, the defenders of the present system could hide behind the pseudo-scientific argument that the social inequality that condemns the majority of men and women to the rubbish heap was the result of "iron necessity," that it was "all in our genes," just as in the past it was all "written in the stars." No longer! The criminal injustice of class society now stands condemned at the court of the very science whose aid it attempted to enlist.

The implications of this are truly staggering. In the course of human history, there have not been many geniuses. It is clear that Albert Einstein had the (genetic) potential to become a world-famous scientist. But it is equally clear that the same Albert Einstein, if born in a Glasgow slum or a village in Ethiopia, would never have become such. The potential would have existed as a bare possibility, but would have simply been wasted. And such is the fate of a very large number of potential Einsteins, Darwins and Beethovens, whose potential is crushed and wasted by this infamous system of capitalism. This terrible waste of human talent has long been reflected upon by the finest minds and noblest spirits. In the 18th century the English poet Grey wrote in his celebrated *Elegy Written in a Churchyard*:

> *Full many a gem of purest ray serene*
> *The dark, unfathom'd caves of ocean bear.*
> *Full many a flower is born to blush unseen*
> *And waste its sweetness on the desert air.*

Trotsky put the same idea less poetically but quite effectively when he asked: "How many Aristotles are herding swine? And how many swineherds are sitting on thrones?" Now this is a very good question, and one which until now the defenders of the established order answered — using the pseudo-scientific arguments so generously supplied from the genetics departments of American (and English) universities: it's all in our genes. It was like a "scientific" version of the old English religious hymn, All Things Bright and Beautiful, from which we used to sing:

> *The rich man in his castle; the poor man at his gate;*
> *He made the high and lowly and ordered their Estate.*

Though men and women have frequently questioned the injustice of class society, their voices have always been drowned by the chorus of voices of the defenders of the status quo, who had a vested interested in demonstrating that this was the natural order of things. Once they affirmed that it was the Will of Gods. Then they said that the slaves lacked an immortal soul. Later they argued that Absolute Monarchy was the product of an inevitable and divinely inspired Order. When driven from these positions, they finally took refuge behind a screen of pseudo-scientific arguments allegedly derived from genetics. Now all this has been blown to smithereens. The difference between the rich man in his castle and the poor man at his gate, in terms of their human potential, is negligible. The difference is not that they are born with different genes, but that one is born into a world of riches and privilege, and given every incentive and opportunity to develop whatever potential he or she possesses, while the other is driven down by poverty and despair. Human potential is ground down just as surely as a seed that is crushed under the heel of a boot.

According to the Celera team, of the two or three billion DNA letters that make up our genes, only 10,000 of them account for any difference between any two individuals. "Really we are just identical twins," says Venter. "But like all twins and brothers and sisters, we are all really different in the way we respond to the environment." The implications are perfectly clear. By changing the material conditions of existence, we can create a favorable environment in which every individual can develop their personal potential to the full. This would mean a new Renaissance — a literal rebirth of humanity — on a far higher plane than anything seen hitherto. That, and nothing less, is the real meaning of socialism. In *Reason in Revolt* we wrote the following:

> The potential of a human brain is limitless. To allow a human being to fulfill this potential is the task of society. Environmental facts can greatly restrict potential or enhance it. Bring up children in bad social conditions, and they will be disadvantaged in comparison with those brought up with all their needs provided. Social background is extremely important. If you change the environment, you change the child. Despite the claims of the biological determinists, intelligence is not fixed or genetically predetermined.

Marx explained long ago that "social being determines consciousness." So-called human nature is not something fixed and immutable. In fact, it has changed many times in the course of millions of years of human evolution. The idea that evolution has reached an end, that men and women have already reached the pinnacle of their physical and mental powers will not be accepted by any minimally cultured person with the slightest knowledge of how our species has struggled to reach the present point of its development. Far from ending, as Francis Fukuyama has suggested, human history has not yet begun. Nor will it begin until men and women finally take their destinies consciously into their own hands.

Ancient Greek mythology has handed down to us the story of Tantalus, the giant who was condemned by Zeus to suffer the torments of hunger and thirst while an abundance of food and drink lay just beyond his grasp. In this myth we have a direct analogy with capitalist society in the period of its decadence. All the material means exist for achieving the goal of socialism — a classless society in which humans will control their own lives instead of being the blind objects of unseen forces beyond their control or understanding. The next great step on the role of human evolution demands that we finally put an end to the degrading social apartheid of class society, that we put an end to the modern equivalent of slavery, and replace capitalist anarchy and the law of the jungle with genuinely human relations. Once we create the necessary conditions for human development, freeing the vast potential that exists in industry, agriculture, science and technology — and above all the virtually infinite potential for development that slumbers inside every human being — the sky would be the limit.

By Alan Woods
London February 16, 2001
(originally published at www.marxist.com)

APPENDIX II

A TRIBUTE TO A GREAT SCIENTIST, STEPHEN JAY GOULD, WHO DIED ON MAY 20, 2002

Sadly, on Monday, May 20, Stephen Jay Gould, the famous American paleontologist, died of cancer. He was 60 and died at his home in New York.

Gould made a major contribution to development of modern science with his theories on evolution. Prior to his studies scientists had accepted Darwin's view of a very slow and gradual process of evolution. Together with Eldredge in the early 1970s, beginning with a study of land snail shells, he discovered that there was another pattern to the evolutionary process. They saw that what the fossil records showed was not one continuous gradual process, but a series of sudden bursts of change followed by relatively long periods of very slow development. Gould and Eldredge coined the term "punctuated equilibria" to describe this process.

Gould and Eldredge then had to face quite widespread opposition from the scientific community. By patiently explaining their ideas they managed to convince many, but some to this day still reject their theory.

Alan Woods and Ted Grant wrote their book, *Reason in Revolt, Marxist Philosophy and Modern Science* in 1995. The book was published 100 years after the death of Frederick Engels and the purpose was to update Engels' *Dialectics of Nature*. The last 100 years of scientific study have provided ample proof that the method of Engels and Marx, i.e. that of dialectical materialism, actually reflects the real processes of nature.

Stephen Jay Gould and Niles Eldredge demonstrated quite conclusively that these processes are sometimes slow and protracted and at other times extremely rapid. They show how a gradual accumulation of small changes at a certain point provokes a qualitative change.

In this they finally resolved the problem Darwin had in understanding what was known as the "Cambrian explosion." Prior to the Cambrian explosion very few fossil records have been found. Then "suddenly" life forms seem to develop very rapidly. In fact fossil records do not fit in to a gradualist interpretation of evolution. There are periods where life forms change rapidly, and then there are other long periods where nothing seems to change. Darwin believed that it was just a question of time before new discoveries in the fossil records would show the gradual change that had taken place. But these records have never materialized. Gould and Eldredge understood what had actually happened.

This discovery was yet another confirmation of one of the fundamental laws of dialectics, the transformation of quantity into quality. Individual, almost imperceptible, small changes pile up one upon another. At a certain point the quantity provokes a sudden leap, a qualitative change. This has been confirmed over and over again in the natural world. It can also be applied to the development of society itself. Small changes over long periods of time suddenly lead to abrupt leaps.

However, the scientific-academic world is dominated by bourgeois ideology, that is, the way of thinking of the capitalist class. The capitalists cannot accept the idea that the natural world and society change through sudden leaps, i.e. revolutions. They want us to believe that everything is gradual. Thus they are imposing upon the real objective processes of nature, and of society, their own subjective viewpoint which is determined by their privileged position. Unfortunately for them the most advanced scientific research repeatedly contradicts this viewpoint. That explains why Gould and Eldredge faced such vehement opposition to the conclusions they drew from their studies.

Gould himself was actually aware of Marxist philosophical thought. In his book *Ever Since Darwin*, he refers to Engels' essay *The Part Played by Labor in the Transition from Ape to Man* and he says the following:

> Indeed, the nineteenth century produced a brilliant exposé from a source that will no doubt surprise most readers — Frederick Engels. (A bit of reflection should diminish surprise. Engels had a keen interest in the natural sciences and sought to base his general philosophy of dialectical materialism upon a "positive" foundation. He did not live to complete his "dialectics of nature," but he included long commentaries on science in such treatises as the *Anti-Dühring*.) In 1876, Engels wrote an essay entitled, *The Part Played by Labor in the Transition from Ape to Man*. It was published posthumously in 1896 and, unfortunately, had no visible impact upon Western science.
>
> Engels considers three essential features of human evolution: speech, a large brain, and upright posture. He argues that the first step must have been a descent from the trees with subsequent evolution to upright posture by our ground-dwelling ancestors. "These apes when moving on level ground began to drop the habit of using their hands and to adopt a more and more erect gait. This was the decisive step in the transition from ape to man." Upright posture freed the hand for using tools (labor, in Engels' terminology); increased intelligence and speech came later.

Gould understood the limitations of Western thought when he wrote that a "deeply rooted bias of Western thought predisposes us to look for continuity and gradual change."

Although the Soviet Union was a terribly deformed caricature of what genuine socialism should be, among Soviet scientists there was a greater understanding of dialectics. And in *The Panda's Thumb* he points out that: "In the

Soviet Union, for example, scientists are trained with a very different philosophy of change — the so-called dialectical laws, reformulated by Engels from Hegel's philosophy. The dialectical laws are explicitly punctuational. They speak, for example, of the 'transformation of quantity into quality.' This may sound like mumbo jumbo, but it suggests that change occurs in large leaps following a slow accumulation of stresses that a system resists until it reaches the breaking point. Heat water and it eventually boils. Oppress the workers more and more and bring on the revolution. Eldredge and I were fascinated to learn that many Russian paleontologists support a model similar to our punctuated equilibria."

Gould was not prepared to go all the way and accept that dialectics can be applied not only to science, and paleontology in particular, but to society itself. Like many scientists he used the dialectical method in his own sphere of studies without grasping the overall outlook of Marxism. However, through his studies he made a major contribution to the development of human thought and of our understanding of the world we live in. Above all he provided more scientific evidence that strengthens the position of Marxism, for it proves that dialectical materialism is not a fantastic notion thought up by Marx himself, but it is simply the reflection of the real material world as it is.

Through his works Gould became one of the most well-known American scientists. He wrote volumes and openly expressed his views in opposition to many gradualist evolutionary theorists. He also popularized his ideas and made them easily accessible to millions of readers. He wrote a long series of essays in *Natural History* magazine. He won the National Book Critics Award in 1982 and came 24th in the Modern Library's list of the one hundred non-fictional English language works of the 20th century. His works were always permeated with a progressive outlook. He totally rejected reactionary scientific theories. For example he refuted all attempts to use pseudo-scientific theories to justify racism and discrimination.

His presence will be greatly missed by all thinking people. We recommend that all our readers take the time to read at least his main works. This is the best tribute we can make to one of the great scientists of the 20th century.

Fred Weston

London

May 22, 2002

(originally published at www.marxist.com)

BIBLIOGRAPHY

Aristotle, *Metaphysics*, London, 1961

Asimov, I., *New Guide to Science*, London, 1987

Barrow, J. D., *The Origin of the Universe*, London, 1994

Berkeley, G., *The Principles of Human Knowledge*, London & Glasgow, 1962

Bernal, J. D., *The Origin of Life*

Bernal, J. D., *Science in History*, London, 1954

Blackmore and Page, *Evolution: The Great Debate*

Bohm, D., *Causality and Chance in Modern Physics*, London, 1984

Bruner, J. S., *Beyond the Information Given*, London, 1974

Bruner, J. S. and Haste, H. (eds), *Making Sense*, London & New York, 1987

Buchsbaum, R., *Animals Without Backbones*, 2 vols., London, 1966

Bukharin, N. I. and others, *Marxism and Modern Thought*, London, 1935

Burn, A. R., *The Pelican History of Greece*, London, 1966

Calder, N., *Einstein's Universe*, London, 1986

Caudwell, C., *The Crisis in Physics*, London, 1949

Childe, V. G., *Man Makes Himself*, London, 1965

Childe, V. G., *What Happened in History*, London, 1965

Chomsky, N., *Language and Mind*, New York, 1972

Cohen and Nagel, *An Introduction to Logic and the Scientific Method*, London, 1972

Cornforth, M., *The Open Philosophy and The Open Society*, London

Cornforth, M., *Dialectical Materialism, an Introduction*, London, 1974

Darwin, C., *The Origin of Species*, London, 1929

Davies, P., *The Last Three Minutes*, London, 1994

Dawkins, R., *The Extended Phenotype*

Dawkins, R., *The Selfish Gene*, Oxford, 1976

Dietzgen, J., *Philosophical Essays*, Chicago, 1917

Dietzgen, J., *The Positive Outcome of Philosophy*, Chicago, 1906

Dobzhansky, T., *Mankind Evolving*, New York, 1962

Donaldson, M., *Children's Minds*, London, 1978

Donaldson, M., *Making Sense*

Engels, F., *The Dialectics of Nature*, Moscow, 1954

Engels, F., *Anti-Dühring*, Peking, 1976 (See also under Marx)

Farrington, B., *Greek Science*, London, 1963

Farrington, B., *What Darwin Really Said*, London, 1969

Ferris, T., *The World Treasury of Physics, Astronomy and Mathematics*, Boston, 1991

Feuerbach, L., *The Essence of Christianity*, New York, 1957

Feynman, R. P., *Lectures on Physics*, London, 1969

Forbes, R. J. and Dijksterhuis, E. J., *A History of Science and Technology, Vol. 1*, London, 1963

Frazer, Sir J., *The Golden Bough*, London, 1959

Freud, S., *The Psychopathology of Everyday Life*, London, 1960

Galbraith, J. K., *The Culture of Contentment*, London, 1992

Gleick, J., *Chaos, Making a New Science*, New York, 1988

Gould, S. J., *An Urchin in the Storm*, London, 1987

Gould, S. J., *Ever Since Darwin*, London, 1977

Gould, S. J., *The Panda's Thumb*, London, 1980

Gould, S. J., *Wonderful Life*, London, 1990

Haldane, J. B. S., *The Marxist Philosophy and the Sciences*, London, 1938

Hawking, S., *A Brief History of Time, From the Big Bang to Black Holes*, London, 1994

Hegel, G. W. F., *The Science of Logic*, 2 vols., London, 1961

Hegel, G. W. F., *Logic, Part 1 of the Encyclopaedia of the Philosophical Sciences*, Oxford, 1978

Hegel, G. W. F., *The Phenomenology of Mind*, London, 1961

Hegel, G. W. F., *Philosophy of Right*, Oxford, 1942

Hegel, G. W. F., *Lectures of the History of Philosophy*, 3 vols., London

Hobbes, T., *Leviathan*, London, 1962

Hoffmann, B., *The Strange Story of the Quantum*, London, 1963

Hooper, A., *Makers of Mathematics*, London

Huxley, J., *Evolution in Action*, London, 1963

Huizinga, J., *The Waning of the Middle Ages*, London, 1972

Ilyenkov, E. V., *Dialectical Logic*, Moscow, 1977

Johanson, D. C. and Edey, M. A., *Lucy, The Beginnings of Humankind*, London, 1981

Johnson, P., *Ireland, a Concise History*

Kant, I., *Critique of Pure Reason*, London, 1959

Kline, M., *Mathematics, the Loss of Certainty*, London, 1980

Kneale, W. and Kneale, M., *The Development of Logic*, Oxford, 1962

Landau, L. D. and Rumer, G. B., *What is Relativity?*, Edinburgh & London, 1964

Leakey, R., *The Origin of Humankind*, London, 1994

Lefebvre, H., *Lógica formal, Lógica dialéctica*, Madrid, 1972

Lenin, V. I., Collected Works, Moscow, 1961

Lerner, E. J., *The Big Bang Never Happened*, London & Sydney, 1992

Lewin, R., *Complexity, Life at the Edge of Chaos*, London, 1993

Luce, A. A., Logic, London, 1966

Lucretius, T., *The Nature of the Universe*, London, 1952

Lukacs, G., *History and Class Consciousness*, London, 1971

Marx, K., *Capital, Vol. 1*, Moscow, 1961

Marx, K., *Grundrissé*, London, 1973

Marx, K. and Engels, F., *Selected Correspondence*, Moscow, 1965

Marx, K. and Engels, F., *Collected Works*, Moscow

Marx, K. and Engels, F., *Selected Works*, 3 Vols, Moscow

Oparin, A. I., *The Origin of Life on Earth*, 1959

Piaget, J., *The Mental Development of the Child*

Plekhanov, G., *The Development of the Monist View of History*

Plekhanov, G., *Selected Philosophical Works*, Moscow, 1976

Popper, K., *Unended Quest*, Glasgow, 1982

Prigogine, I. and Stengers, I., *Order Out of Chaos, Man's New Dialogue with Nature*, London, 1985

Rees-Mogg, W. and Davidson, J., *The Great Reckoning — How the World Will Change in the Depression of the 1990s*, London, 1992

Rhodes, F. H. T., *The Evolution of Life*, London, 1962

Romer, A. S., *Man and the Vertebrates*, 2 vols., London, 1970

Rose, S. and Appignanesi, L. (eds), *Science and Beyond*, Oxford, 1986

Rose, S., Kamin, L. J. and Lewontin, R. C., *Not in Our Genes*, London, 1984

Rose, S., *The Conscious Brain*, London, 1976

Rose, S., *The Making of Memory*

Rose, S., *Molecules and Minds*

Savage-Rumbaugh, S. and Lewin, R., *Kanzi — The Ape at the Brink of the Human Mind*, London, 1994

Stepanova, Y., *Frederick Engels*, Moscow, 1958

Spinoza, *Ethics*, London, 1993

Sutherland, P., *Cognitive Development Today: Piaget and his Critics*, London, 1992

Stewart, I., *Does God Play Dice?* London, 1990

Toulmin, S. and Goodfield, J., *The Fabric of the Heavens*, London, 1961

Trotsky, L. D., *In Defence of Marxism*, London, 1982

Trotsky, L. D., *My Life*, New York, 1960

Trotsky, L. D., *Literature and Revolution*, Michigan, 1960

Trotsky, L. D., *The Living Thoughts of Karl Marx*, New York, 1963

Trotsky, L. D., *Their Morals and Ours*

Trotsky, L. D., *Problems of Everyday Life*, New York, 1979

Trotsky, L. D., *The Struggle Against Fascism in Germany*, New York, 1971

Trotsky, L. D., *Writings, 1939-1940*, New York, 1973

Waldrop, M. M., *Complexity*, London, 1992

Walter, W. G., *The Living Brain*, London, 1963

Washburn and Moore, *Ape To Man: A Study of Human Evolution*

Westbroek, P., *Life as a Geological Force*

White, M. and Gribbin, J., *Einstein, A Life in Science*, London, 1993

Whitehead, A. N., Adventures of Ideas, London, 1942

Wills, C., *The Runaway Brain, The Evolution of Human Uniqueness*

Wilson, E. O., *Sociobiology – The New Synthesis*, Cambridge, 1975

GLOSSARY OF TERMS

Please note that this glossary is not intended, for reasons of space, to be exhaustive. To avoid repetition, terms explained in the text are not generally included here.

Adaptive radiation - Evolution, from a primitive type of organism, of several divergent forms adapted to distinct modes of life.

Allopatric Theory - The theory that the evolutionary divergence of populations into separate species, which no longer interbreed, takes place in geographically separate places

Amino Acids - Organic compound containing both basic amino and acidic carboxyl groups. Amino acid molecules combine to make protein molecules and are therefore a fundamental constituent of living matter.

Causality, Law of - The law defining the interdependence of cause and effect - the necessary connections between phenomena. Causality is an essential question in the struggle between materialism and idealism.

Chromosomes - A chain of genes found in cells. They are present in all cells in the body and consist of DNA and a supporting structure of protein.

Cognition - The process by which human thought reflects and observes the real world.

Convergent Series - Number series in which the successive partial sums obtained by taking more and more terms approach some fixed number or limit.

Cytoplasm - All the protoplasm of a cell excluding the nucleus.

Determinism - A belief that all processes are predetermined by definite causes and natural laws and can therefore be predicted. Biological determinism and mechanical determinism are two variations of this premise. Indeterminism is the reverse of this - a belief that events are governed not by laws but by pure chance.

Dialectics - From the Greek words for dispute and debate, this is the science of the general laws governing the development of nature, science, society and thought. It considers all phenomena to be in movement and in perpetual change. Marxism linked this concept to materialism and showed the process of development in all things through struggle, contradiction and the replacement on one form by another.

Diploid - Cell with chromosomes in pairs. DNA - The molecule that carries the genetic information in organisms (except RNA viruses).

Dogma - A blind belief in things often without a material base.

Eclecticism - A mechanical and/or arbitrary collecting of concepts or facts without any pre-established principles or structures. Eclecticism is often used to attempt to reconcile the irreconcilable such as idealism and materialism.

Electrons - Elementary particles that possess one unit of negative charge and are a constituent of all atoms.

Entropy - One of the main notions of thermodynamics, where it is normally viewed as a measure of disorder. In isolated systems, it is used to determine the way in which the system will change if heated or cooled, compressed or expanded. Thermodynamics holds that the entropy of a system can never decrease but only increase and that a state of maximum entropy is marked by a state of balance in which no further conversion of energy is possible. This has been used to justify the erroneous idea of the "heat death of the universe." In recent years, I. Prigogine has reinterpreted the Second Law of Thermodynamics in a way which defines entropy differently. According to Prigogine, entropy does not mean higher disorder in the generally accepted sense, but an irreversible process of change which generally leads to more highly ordered states.

Empiricism - A teaching on the theory of knowledge which holds that sensory experience is the only source of knowledge and affirms that all knowledge is founded on experience and is obtained through experience. The opposite to rationalism. The main failing of this is a tendency to reject reason as a means of deduction in favor of a metaphysical exaggeration of the role of experience alone.

Electromagnetism - The study of the effects of the relationship and interplay between a magnetic field and an electric current. For example the electrical creation of a magnetic field in a conductor.

Eugenics - A doctrine which holds that the human race can be "improved" by selective control of breeding to eradicate less "desirable" traits in society. The supporters of eugenics argue that social problems are caused by inherited genetic traits in people which can be bred out to resolve the problem for future generations. The logical conclusion of this theory is deeply racist and reactionary based on dubious research and prejudice.

Eukaryotes - One of the two major groups of organisms on Earth (the other being Prokaryotes). Characterized by the possession of a cell nucleus and other membrane-bounded cell organelles.

Gene - A unit of heredity; a sequence of base pairs in a DNA molecule that contains information for the construction of protein molecule.

Genome - The entire collection of genes possessed by one organism.

Genotype - Genetic constitution (the particular set of alleles present in each cell of an organism) as contrasted with the characteristics manifested by the organism.

Gradualism - The theory that all evolutionary change is gradual rather than occurring in leaps and jumps.

Haploid - Cell with single set of chromosomes.

Lamarckism - The theory that acquired characteristics can be inherited and that any new genetic variation tends to be adaptively directed rather than 'random' as stated by Darwin.

Logical Positivists - A variation on positivism which attempts to combine subjective-idealist empiricism with a method of logical analysis.

Lysenkoism - A revival of Lamarckism in the USSR under Lysenko who sought to affect the hereditary modification of plants by certain treatments. his research was subsequently discredited but was heavily touted by Stalinists in its day.

Malthusian Theory - The theory developed by Thomas Malthus which claimed that population levels were responsible for social problems and should be checked to resolve them since uncontrolled population increases occur on a geometrical ratio whereas the increase in resources occurs on an arithmetical basis. This is not so but laid the basis for the belief that nothing could be done about the problems of the world. In its most extreme form it was the basis for an acceptance of famines etc. as unavoidable and socially necessary.

Meiosis - Cell division in which a cell gives rise to daughter cells with half as many chromosomes.

Metaphysics - There are two definitions of this word: the one used by Marx and Engels, and the other more traditional conception. In Marxist terminology, metaphysics is a method which holds that things are final and immutable, independent of one another and denies that inherent contradictions are the source of the development of nature and society but rather that nature is at rest, unchanging and static. All things can be investigated as separate from each other. Nowadays, the word reductionism would often be used instead.

The more traditional philosophical definition derives from Aristotle who used the word metaphysics to describe the branch of philosophy dealing with universal concepts as opposed to the observation of nature (in Greek, "meta ta physika" means "that which comes after physics"). Later on it became a synonym for abstract idealist speculation.

Mitosis - Cell division in which a cell gives rise to daughter cells with a complete set of chromosomes.

Monad - A primary organic unit. A chemical element having a valency of one. The monad played a central role in the idealist philosophy of Leibniz.

Mutation - An inherited change in the genetic material; a change in the genotype

Neutrons - One of the two types of particle which form the nucleus of an atom - the other being the proton.

Nodes - The points in a wave system where the amplitude of the wave is zero. In Hegel, the nodal line of measurement was one where the line is interrupted by sudden leaps, denoting qualitative change ("node" here means "knot").

Nucleotide - A biochemical molecule used as the basic building block of DNA and RNA.

Paleontology - The study of fossils and other records of ancient life.

Phenotype - Manifested attributes of an organism (e.g., eye color).

Photon - Units or 'packets' of electromagnetic radiation.

Plasma - A gas that contains a large number of positively and negatively charged particles (ions and electrons). This can occur when a gas is raised to extremely high temperatures (e.g., the outer regions of the sun) or in an intense electrical field. Plasma physics is an important branch of modern science.

Polymorphism - The coexistence of several well-defined distinct phenotypes or alleles in a population.

Positrons - The antiparticles of electrons - having the same mass but a positive charge.

Positivism - An idealistic current which believes in "positive" facts rather than abstract deductions. It denies that philosophy is a world outlook and states that belief should be concentrated on a description of facts rather than an analysis of them. Positivism claims to be neutral and above philosophical outlooks, interested in processes but not willing to go beyond the boundaries of the status quo. In effect they confirm the maintenance of existing social structures.

Prokaryotes - One of the two major groups of organisms on Earth (the other being Eukaryotes). They have no structured cell nucleus and no membrane-bounded organelles.

Proton - One of the two types of particles which form the nucleus of an atom - the other being neutrons.

Protoplasm - Substance within and including plasma-membrane of a cell or protoplasm.

Quantum Mechanics - The mathematical description of the workings of the atomic and sub-atomic structures.

Quark - According to particle physics these sub-atomic particles are believed to be the constituents of elementary particles known as hadrons. Five or possibly six different sorts are thought to exist, but new discoveries are being made all the time.

Quasars - Quasi-stellar radio sources (quasars) were first detected by virtue of their radio transmissions and appear to show the small bright centers of distant galaxies (although some believe that they are not as far away as people imagine but are moving at high speeds).

Rationalism - The theory which holds that reason is the unique source of knowledge as against empiricism which holds that perception is the source of knowledge.

Reductionism - A belief that all scientific laws and processes relating to complex systems can be reduced down to basic scientific laws. Physicalism was a version of this.

Relativity, Theory of - The laws of relativity (relationship between an object and an observer or another object) considered and developed by Einstein. Einstein's general theory deals with motion, gravity, time and the concept of curved space. The theory which deals with constant velocities is called the special theory. The most famous part of these laws is that which shows the relationship between mass and energy ($E = mc^2$).

Speciation - The process of evolutionary divergence i.e., two species being produced from one source.

Stasis - A period in which no evolutionary change takes place in the development of a species.

Sufficient Reason, Law of - A principle that holds that a proposition can only be considered true if sufficient reason for it can be formulated.

Syllogism - A doctrine of inference, historically the first logical system of deduction, formulated by Aristotle. Every syllogism consists of a triad of propositions: two premises and a conclusion.

Systematics - Study of the diversity of organisms.

Taxonomy - Study of classifying organisms.

Thermodynamics - The branch of physics concerned with the nature of heat and its transformations. The First Law of Thermodynamics is generally referred to as the Law of the Conservation of Energy. The Second Law deals with the concept of increasing entropy (see under entropy).

Made in the USA
Monee, IL
07 July 2026

56551552R00148